"When billions of things are connected, talking and learning, the only limitation left will be our own imaginations."

— CISCO, 2011

BIS Publishers
Building Het Sieraad
Postjesweg 1
1057 DT Amsterdam
The Netherlands
T +31 (0)20 515 02 30
F +31 (0)20 515 02 39
bis@bispublishers.nl
www.bispublishers.nl

Designed by: Booreiland
www.booreiland.nl

Edited by: Kumar Jamdagni
www.language-matters.nl

ISBN 978 90 6369 251 3

SARA CÓRDOBA RUBINO WIMER HAZENBERG MENNO HUISMAN

BIS PUBLISHERS

SARA CÓRDOBA RUBINO

Sara was born in Mexico City and did a bachelor in Industrial Design at the Universidad Autónoma Metropolitana. Afterwards she applied her skills as industrial designer at the Papalote Children's Museum, by working on various interactive exhibitions. In 2007 however, Sara became account manager for Cajaplax S.A. de C.V., a packaging design and production company. Her activities varied from creating various packaging concepts, overlooking the overall design process and managing the company's accounts such as Unilever, GlaxoSmithKline, Bayer and Abbott Laboratories.

Having the urge to explore beyond Mexican borders, Sara flew off to The Netherlands in 2008 and did a master's degree in Strategic Product Design at the faculty of Industrial Design Engineering at Delft University of Technology. This programme aims at the strategic stages that precede the actual new product and service development phases. Now, 2011, Sara applies these skills at Dutch design studio Booreiland as researcher and project manager. She is responsible for the foundation of Network Focused Design, a design approach Booreiland uses for its projects.

In her spare time, Sara likes to practice dancing and boxing, not an everyday combination.

WIMER HAZENBERG

Wimer was born in Utrecht, The Netherlands, but soon moved to the upper regions of the country. After 1 year of art school, Wimer decided to do something else and received his master's degree in Artificial Intelligence at the University of Groningen. Ever since, he has been hooked on AI programming experiments. Some of these experiments are collected on Monokai.nl, his online playground of creative output. Wimer has also created Monoslideshow, a best-selling ultra customizable photo slideshow module web developers can embed in their website.

In 2003 Wimer co-founded Booreiland, a design studio specialized in design and strategy for web and print projects. Within the company Wimer works as creative director, running both commercial and internal projects. In 2008 he was responsible for launching the internal project Vormator, a design contest in which artists had to create a graphic with only eight shapes and a set of given rules. This contest proved that even when artists are heavily constrained, it has little impact on creativity or variety of the resulting artworks. The best creations have been published in a book, called Vormator - The Elements of Design.

In his spare time, Wimer makes electronic music and still is determined to release a music album one day.

MENNO HUISMAN

Menno was born in Leeuwarden, The Netherlands, but took off to Delft when starting his studies in Industrial Design Engineering at Delft University of Technology. He received his master's degree in Integrated Product Design, which focuses on the design and engineering facets of consumer products. However, Menno has always been fascinated by web and graphic design as well, and therefore always looks for total visual brand experiences in his creations.

When he co-founded design studio Booreiland in 2003, he incorporated his concepting and design methodology skills from his background in Industrial Design in the company. He discovered graphic and web design disciplines often lack structured approaches, and introduced design tools to ensure a continuous stream of high quality work coming from Booreiland. Menno currently works as creative director, running both commercial and internal projects. In 2009 he was responsible for launching internal project Okimok, a photo sharing web service that is a great showcase for Booreiland when it comes to the studio's design and development skills.

In his spare time, Menno likes to work on his BMW E21, which needs love and continuous maintenance.

With their backgrounds in artifical intelligence and product design, Wimer and Menno soon discovered a growing trend of products being connected to the web in new and smart ways. By that time Booreiland had already created lots of web experiences and digital interfaces for its clients, so it was only a small step to start up an internal research project to investigate this new field, later to be named Meta Products. With Sara joining the company, Network Focused Design was laid out as a design approach for Booreiland to design Meta Products.

With this book the authors not only write about their vision on Meta Products, they also present Network Focused Design (licensed under Creative Commons) in a much broader setup, usable for everyone who wants to design successful solutions for our connected world.

Sara, Wimer and Menno currently live and work in Amsterdam. Great city, visit it when you have the chance!

Booreiland has worked for clients such as Philips, South Africa Tourism, Wacom Europe, ABN AMRO, KLM, Mexx, BIS Publishers, BiD Network and Stibbe.

Have a look at the studio's work on www.booreiland.nl

FOREWORD BY MIKE KUNIAVSKY

PHOTO © JAMES HOME

Mike Kuniavsky is a designer, writer, researcher, consultant and entrepreneur focused on people's relationship to digital technology. He cofounded Adaptive Path, an influential San Francisco design consulting firm, and ThingM, a ubiquitous computing design studio and micro-manufacturer. He is the author of 'Observing the User Experience,' a popular textbook of user research methods, and 'Smart Things: ubiquitous computing user experience design,' a guide to the user-centred design of digital products.

Designers must regularly walk the space between broad theories and tiny details, without losing sight of either one. This is especially true of design for the new environment of technological experiences, experiences that exist simultaneously in physical lives and in the Cloud world of internet services — the space that this book calls Meta Products. A car borrowed through a car sharing service can take you from place to place and is simultaneously a data object manipulated by multiple digital services.

Creating experiences in this ecology of people, products and services challenges the still young discipline of interaction design. Until recently, only a few guides even acknowledged that what's important in digital networked object design is not so much the object, but the social relationships around it.

This book is one of the first to address this challenge directly. It provides the context, history and key ideas of Meta Product design using great quotations and witty illustrations. The authors take a practical approach to designing in this new environment, recommending specific tools and techniques. Never rigid or prescriptive, their advice applies to designers in start-ups and large companies alike. They argue strongly for mindful design and reflective designers, but never for a single specific process or sole technique. Most importantly, they never promise any magic or single path — there can't be at this early stage of a change that may be on the scale of the Industrial Revolution — but begin a conversation.

It's a conversation that is going to last for decades to come, and there are few better places to start.

INTRODUCTION

Imagine yourself opening a door and suddenly finding yourself floating in space and you see the whole world in front of you. All of it, in all its infinity. Every mountain, every animal, every river, every culture. You can see everything and nothing is hidden from view. You could spin it with your fingertips and zoom in and out, rescaling the world as you wish. As close by as your kitchen, or as far away as the remote Marshall Islands (latitude 9°00, longitude 168°00). You could jump in and out of it wherever you want. Where would you like to go? What would you like to see? What would you like to do?

This is what our current society and the ever-improving web and ubiquitous technologies are making possible. Our world is in a transition period where social change will no longer be dictated by the centralized processes of industrialization, but by information. The information that is generated and enabled by everything we do in the search for relevant alternatives, efficiency, authenticity, immediacy and many other aspirations. While we enjoy the fruits of industrialization, but struggle with them at the same time, the web and ubiquitous technologies are making it possible to add a layer of connectedness to objects, products, services and environments. We will be able to sense, track, measure and control information in limitless forms. This information will fuel up our lives, and hopefully help us improve them. Finding smart ways to use this fuel can be an overwhelming challenge for many of us, particularly for designers. Bits and atoms have never been so interrelated as they are today, given the potential that designers have to embed smartness, enable connectedness and power information streams in everything they design, either at the object level or the environment level. Consequently, potential new meanings, new interactions, new ways of building knowledge will arise, and ultimately new behaviours. We believe designers are key to helping people and organizations make sense of this complexity.

Designers are by nature good in dealing with complex and undefined situations, but we have noticed that there is little formal literature acknowledging that what the future is bringing, both socially and technologically, requires a shift in some design approaches. In the near future, we will be seeing more and more services incorporating real-time remote activities, connectivity on demand, multi-users, parallel tasks, devices talking to people and to other devices, sometimes automated, sometimes controlled — all in dedicated networks we like to call Meta Products. We believe that now is the time for designers to reflect and gear up for the near future.

Structure of the book

This book is the result of observations and reflections on our own experiences, fascinating conversations with experts, and a vast amount of reading. We try to provide an extensive overview of the subject while recommending specific practices. Two important buzzwords you will meet throughout the book are: reflective practice and Network Focused Design. Most of the book (the first four chapters) consists of a journey of reflective practice. In these chapters we aim to start a dialogue with you in order to explore the different aspects that lead to innovation and the underlying human dimensions that allow people to adopt new technology.

The first chapter is a reflective journey through the past where we invite you to use your design lens to discover how people's aspirations have led to the greatest changes in history up until today. It also introduces the phenomenon of Meta Products. The second chapter is about the future, and aims to trigger you to build your own way of envisioning the future; to think of the possibilities, the impact of those possibilities, and to reflect on your design role. The third chapter explains the disruptive part of the phenomenon of Meta Products, and the challenges it brings to the design practice. You will find out how designers are required to design networks, work in transdisciplinary teams and investigate how people assign meaning and value to the network of services and products they use. We then go on to define our ideals behind designing Meta Products in chapter four.

The final chapter is dedicated to our design approach, called Network Focused Design. Here we provide you with a more operational perspective to some of the concepts treated in the first four chapters. It's a mix of service, interaction and product design techniques and principles we believe are appropriate to designing Meta Products. The approach is one that will take you out of the technology box and help you focus on people's contexts and aspirations to help you pinpoint possible areas of innovation. In nine simple steps, you will read about recommendations of activities and tools that you can easily adapt to your own design practice. The only requirement on your part will be that you are looking to design interactions that truly serve people. We don't claim or attempt to provide one unique answer, that would be senseless and unrealistic. Our aim is to invite you to reflect on the way you are working, and to customize Network Focused Design to your own needs.

How to use this book

This book has been structured in such a way that you can read it right through starting from chapter one and ending at chapter five. Or you can read it in whichever order you prefer, as the chapters have also been written as independent entities. Each chapter concludes with a summary of the most relevant insights covered in that chapter. You could take a look at them before starting a chapter to get you familiarized with the content.
In general, the book has been written with an academic mindset but in a narrative form. Chapter five presents the Network Focused Design approach and contains steps, techniques and recommendations. Throughout the book you will also find cases of interesting Meta Products. The purpose of these cases is to give you a more practical view on what's actually happening in the field of Meta Products today.

We were very lucky to be able to interview experts in various fields related to Meta Products and this book would not have been possible without the inspiring conversations we had with them and other practitioners. We have included their most interesting insights throughout the main text of the book and a selection of Q&As is also included at the end of every chapter.

We wrote this book with industrial, service, interaction and strategic designers in mind. But you will also find this book useful if you are a creative practitioner, a trend watcher, or a business adviser and you are noticing that your core activities or those of your clients are no longer standalone entities but part of a network of services, products, people and environments; or if you are interested in new ways of building information streams, or if you are just trying to get a grip on the impact that the web and other related technologies will have in the near future. We hope you enjoy reading it, and that you become inspired along the way.

CHAPTER 1 — *The Phenomenon*

change / aspirations / reflective practice / design lens / new interactions / industrial society / information / strategic resource / new journey / industrialization consequences / breaking paradigms / new aspirations / information-fuelled / dedicated networks / physical means / web layer / communication goals / touchpoints / system oriented / people oriented / sensing / tagging / tracking

1.1 HISTORY OF CHANGE

Throughout time we've used different ways to deal with change. Someone in the nineteenth century for instance would probably have responded to change very differently from the way you do. Back then, traditions, religion, national identities and simply distance and time were constraints to all that was new; changes took place slowly. Today's world is totally different, the filters used to constrain cultures from being exposed to external influences are more porous, gaining access to knowledge is easier, and the patterns of time demand that we respond almost instantaneously to the ever increasing stimuli we face. All this makes it difficult for us to realize what exactly is changing today.

In any case, whether centuries ago or today, people's aspirations are the drivers of change. The dictionary definition of aspiration is: a strong desire to have or achieve something. Moreover, our aspirations change as we accomplish change. Change is the end point of fulfilling aspirations and the starting point for building new ones. How exactly these aspirations are created in our minds and in the fibres of our society is a rich recipe with many ingredients. The geographical context and the available resources, the ideologies, the religious fundaments, basic human instincts, serendipity and so on — all these ingredients combined in various ways lead to building our aspirations and are therefore what drive us to change our world. But not everybody's aspirations have led to major change throughout history. We have seen how the impact of the aspirations of emperors, of the clergy, of aristocrats, of the educated classes and, today, of almost everybody, has changed throughout time.

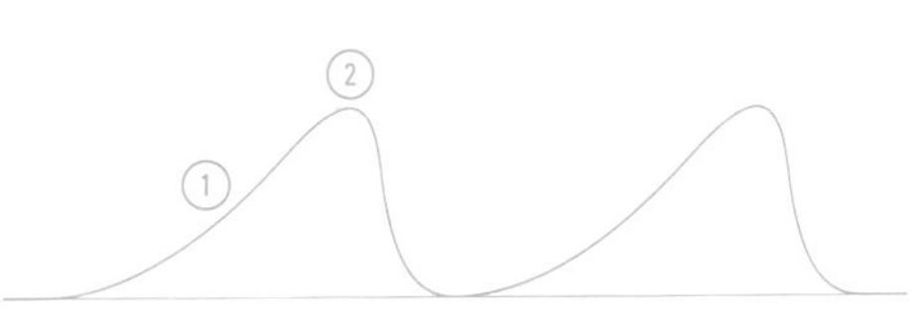

(1) BUILDING ASPIRATIONS DURING TRANSITION PERIOD
(2) FULFILLING ASPIRATIONS AS CHANGE OCCURS

Designers also have an impact on the future as they are constantly looking ahead. If you are one, you probably focus on what motivates people to adopt change. You are constantly identifying aspirations even before people become aware of them. You will be continually busy trying to foresee people's reactions to the things you design while you look at history through a design lens. Historians for instance mostly focus on the methodical narrative and research of what has happened. But you try to identify patterns from the past to understand the present and then discover an opportunity for change (and you can do this very quickly!). Actually you want to 'break' patterns to find new ways of doing things. You use history as a means to find empathy with the present and to help you delineate 'the box' you want to get out of. As a result, the vision you get about the future can help you to question and confront the present so that your intuitive powers are intensified and your synthesizing skills enhanced. You do all this as a reflective practice that enables you to translate people's aspirations, increasing the chance for your design skills to hit an innovation spot.

The main purposes of this chapter are to begin putting into practice your reflective skills and to gain an understanding of how people build aspirations that drive change. You will then be able to define your own criteria about the Meta Product phenomenon. The last part of this chapter contains a more technical, or hands-on, perspective of the phenomenon.

Let's use our design lens to look at some of the major changes that have occurred in history and reflect on the many combinations of ingredients that have led to them. Let's travel through history for the next few pages and try to focus on how human aspirations were intricately built to drive change. For instance, Francis Bacon (1561 - 1626) forecast that natural philosophy (science) could be applied to the solution of all practical problems. Bacon's aspirations were a reflection of the Age of Reason. In his view, machines would liberate mankind and they would save labour, which could then be utilized elsewhere. It was just a matter of time (and serendipity), before scientific effort and intellectual aspirations combined to produce something revolutionary. This happened in 1750, with the patenting of the steam engine. It was not just an invention but a platform of knowledge, ideas and techniques, allowing established industries and new ventures to grow exponentially, affecting people's lives all over the world. In fact, a new idea of progress was born: machines would produce a better society. In fact the aspirations behind the industrial revolution were humanitarian. How wonderful it sounded to be free from all the labour we don't want to carry out... let machines do it!

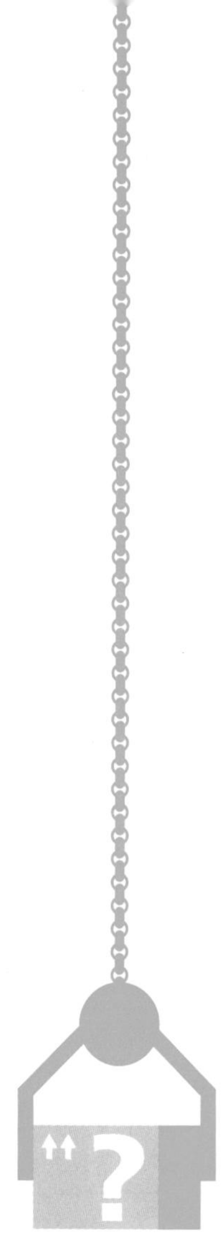

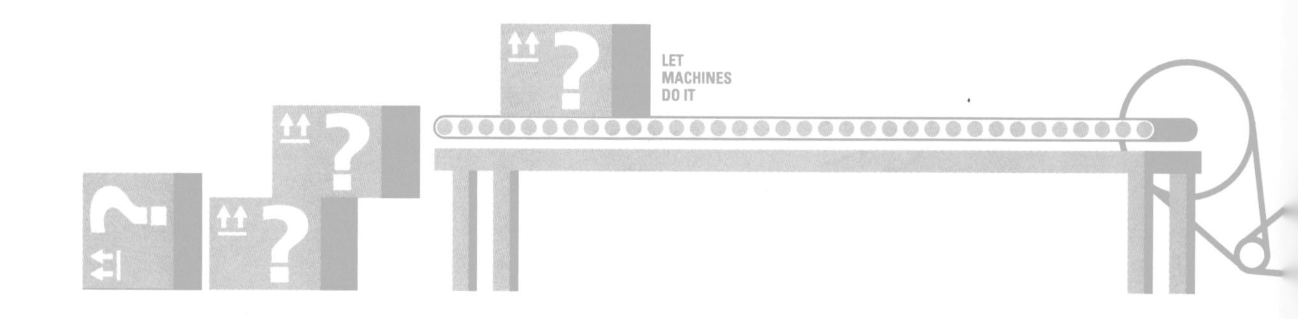

FORD
T-MODEL
ASSEMBLY LINE

This aspiration caused people to look for ways to make the industrial society flourish. One way was by finding faster, easier and standardized production processes. In the heat of pursuing progress at a fast pace, an abundance of products and industrial processes appeared almost everywhere. Industries strove for unlimited economic growth and improvement that provided novel products to people on the one hand, and the first negative repercussions on the other hand, such as poor labour conditions. Now the new products were in the hands of the majority instead of the elites and a new industrial labour force behind these products rendered new social interactions and new ways of learning and building new aspirations.

Can you imagine the unrestrained optimism of the time? How would the world look today without the steam engine?

The mix of ingredients enabled a wider diversity of ideologies than ever before. However, with such a newly born diversity, conflicts between cultures and political geographies arose. Two world wars were sustained through technological capabilities and an old-fashioned militarism, voracious imperialism (hungry for the resources in Africa and parts of Asia) and the nationalistic ideas of the time. After the Second World War, a thriving industrial society continued to reinforce industrial capitalism, feeding the insatiable hunger of progress and the infinite growth of corporations and fast growing populations. In the aftermath of such agitations, people's aspirations for change altered radically; now they had grown aware of the individual's autonomy and the right for truth and happiness. This awareness provoked a much larger explosion of ideologies, particularly in the 1970s. Key movements rose up and flourished, led by a public demanding radical changes in politics and in the distribution of power. But the fast pace of industrial society had already inflicted drastic changes to every aspect of people's lives. Industrialization also enabled and defined relationships between countries. It made people conscious of the concept of global economies and, with this, of the relationship between industrialized countries and the so-called Third World countries (those that were not aligned to the industrialized economies). There was a growing realization of the magnitude of the snowball effects in every corner of the planet. Right at the start of the globalization movement in the 1970s, at a time when Intel introduced the microprocessor, Apollo 17 brought back 250 samples of rock and soil from the moon, Atari produced the first low-priced game console, jumbo jets doubled air passenger capacity, the first test-tube baby was born, the first network email was sent, followed by the first scanned retail barcode, Daniel Bell foresaw that it was no longer the industrial process (at least in industrialized countries) that would dictate social change, but the creation of a service economy in which the technical professionals, information and theoretical knowledge would be the source of innovation, making a self-sustaining technological growth possible. In a world in which almost every single aspect of people's lives was determined by the industrial standardization for mass production and the unlimited use of resources such as oil, Bell was one of the first to express anxiety about air and water pollution, famine, overpopulation, climate change, natural resources and other major global consequences. Back then, Bell saw how knowledge and information would become the strategic resource and the agent of change. And the computer, or any other technology, would be purely instrumental in helping us to deal with the complex interactions he saw coming. At around the same time, there were others, like Weizenbaum, who were worried about the power assigned to the computer and the side effects of the actions based on them[1]. And this was even before the introduction of the first personal computer in 1982. All this conjecture was possible because the United States had gathered all the ingredients together to allow changes to happen faster than anywhere else in the world. On the one hand, the same industrial society with its ideals of progress was politically and financially supporting the development of technologies that would help industries and governments achieve those ideals. On the other hand, in the midst of enormous material abundance, the preoccupation with self-discovery and self-accomplishment became a mass phenomenon, feminist and gay liberation movements flourished, the urge for political openness — as well as many other ingredients resulting from the influences of the industrial processes — were evolving rapidly. Almost imperceptible to the industrialized bureaucratic eye, the internet had already been in the making for a while. It was not one technician's invention but the result of collaboration by many individuals and organizations, originally looking to create ways to deal with information flows in large communication networks. It was in 1991 that the World Wide Web was first introduced to the public. And with this, we witnessed the start of a new society, perhaps one similar to Daniel Bell's vision of post-industrial society: "The post-industrial society... is a 'game between persons,' but a game between persons requires

INTERNET OVERCOMES LIMITATIONS OF DISTANCE, TIME & KNOWLEDGE

increasing amounts of coordination...."[2]. He believed then that the advances in computers and communication would make such coordination more feasible (although of course he had no idea to what extent we would need to coordinate everything today!). He wasn't far wrong, because he believed that the crucial decisions to change are of a sociological nature rather than a technological one.

The introduction of the World Wide Web heralded a new journey. How different were our aspirations now, compared to the age of the industrial society? And whose aspirations were they? One sure thing is that the industrial society's general idea of progress was not able to sustain the new aspirations that had appeared almost unnoticed, but that were growing exponentially as the internet technology advanced, and as people began to be aware of its potential. It was astonishing that anyone (and this time literally anyone) with a computer was able to overcome limitations of distance, time and knowledge to connect with people in far-off countries and in a much more enriching way than any other technology had offered before.

"Technology's greatest benefit is the empowerment of the individual. With smartphones and task-specific software we can instantly make choices about almost anything." — ANNE LISE KJAER

1.2 THE TRANSITION TODAY

The idea of our world as a 'global village' was new and exciting and the internet was enabling a platform where anyone could create and share their own information. In some ways it almost seems like a counter-attack was being launched as a backlash to being enclosed in an archetype of the massive, passive and standardized consumer that simply consumed; the individual now had unique aspirations and was actively pursuing them. Today we are right in the middle of the transition period, and as with most transitions, not everything is running smoothly. On the contrary, today's transition period could be considered as being rather disruptive. There is no doubt that the consequences of industrialization have been enormously constructive but also devastating. From environmental threat, to overpopulation, to the complete reliance on oil just to name a few. We are dealing with the consequences of the consequences, and we will likely still be dealing with them for some time to come.

"We have this world with ridiculous amounts of damage of different kinds and we cannot pretend designers were not implicated in many ways in that. Now we are aware that designers are social actors, we are designing the world through our devices, our products, our services. And that is what I mean with being more in tune with the consequences, to look at the bigger picture and be conscious of the impact of what you're doing." — LUCY KIMBELL

These same corporations and institutions are dedicated to the industrialization of our world and some of us are having a hard time adapting and resisting the consequences. Though small and medium enterprises are proving to be agile and are finding new ways of working.

"SMEs can: 1. react faster 2. create networks of intelligence and get knowledge where they need it 3. prototype and produce things that were very difficult to do years before." — FEDERICO CASALEGNO

On the other hand, we've witnessed how people have adopted the internet and other communication technologies, making possible not only massive technological enhancement but also major social change. Internet has been — or at least started out as — a no-man's land in many senses; free and hard to grasp or delineate, particularly for centralized bureaucratic governments or institutions that view the way things work on that Cloud we call the internet as a rather alien phenomenon. The hacker, for instance, has ignored social etiquette and violated every rule in the book to develop, at zero costs, computing skills that have evolved to become the major opportunities and threats of the internet — and that can be exponentially copied by adopters all over the world.

Today we are building new aspirations, and the ingredients at hand for building them are varied and complex. While many of us might still take for granted the material abundance and infinite growth of our individual interests, an ever-increasing group has transferred this same idea to the world of bits and information. Inevitably, their hunger for information and the possibilities afforded by the available technology so far has helped to fuel new aspirations. We see how greater sections of our society with access to so much information today are looking for truly better alternatives in everything they do. They're looking for fair options, customized products, and relevant information that help them to achieve their individual aspirations. While on the surface this might not seem so revolutionary, it is quite conceivable that a combination of factors will result in them taking over and then breaking down paradigms of ownership, value, relevancy, temporality, globalization, many social structures and institutions such as churches, companies, schools, religious groups, the way we interact with our environment, and so on. How will this happen? In short, by discovering (as technology continues to advance) much more clever ways to use information about potentially everything around us. These ways will aim to fulfil the newly born aspirations and this in turn will break down paradigms on a massive scale.

There is a whole group of interrelated aspirations which will require a powerful and reliable ubiquitous technology and new breeds of designers to serve the individual aspirations of immediacy, local relevancy and the creation of valuable networks. It's precisely in this transition period that your role as a designer becomes crucial. Find out more about the design teams in chapter 4.

For instance, the paradigm of ownership is already changing; owning something will not represent any advantage if what you own is not suited to serving your needs at the moments and places you require them. Wouldn't it be better just 'owning those moments'? This could revolutionize the whole system we have created so carefully for determining the value of products and services. It might even change the way we deal with the concept of money when technology can actually trace 'the right moment and the right place' for you and everybody else.

THE PARADIGM OF OWNERSHIP

1.3 WHAT IS A META PRODUCT?

In the current transition period we are living in, we are building new aspirations that demand increasingly clever ways to use the potentially endless amounts of information we can track, sense, measure, share and produce via the constantly improving ubiquitous technologies.

From both social and technological perspectives, there is a great potential today to revolutionize many aspects of our lives by making our processes ever more efficient and intuitive. It can be medical, logistics, manufacturing, entertainment, food, transportation processes and so on. The transition we are living now is bringing information-fuelled products and services that will be around us as a network whenever we need them. The result of finding (or designing) these increasingly clever ways to use information is what we call Meta Products.

We chose this term because we are making the shift to a society that is no longer determined by the material multiplication of industrialization (product), but by the information generated by our actions (where a product — in the industrialization sense — may not necessarily be involved). Hence 'Meta', which means 'higher or beyond' in Greek. It encapsulates the idea of the quest to transcend to a higher place. But let's not misinterpret the term, because Meta Products are not products that are better or that have more features added to them. Meta Products are dedicated networks of services, products, people and environments fed by the information flows made possible by the web and other ubiquitous technologies. Our working definition of a Meta Product would therefore be: web-enabled product-service networks.

So far, we've explained Meta Products in the socio-cultural context. In the following sections we will describe a more technical perspective so that you get a more hands-on idea of what a Meta Product is.

At the head of these technologies is the web. As the web becomes more and more mobile and robust, it empowers other technologies such as Near Field Communication (NFC), Radio Frequency Identification (RFID), Wireless Sensor Networks (WSN) or Augmented Reality (AR). These technologies are becoming ubiquitous almost in a snap!

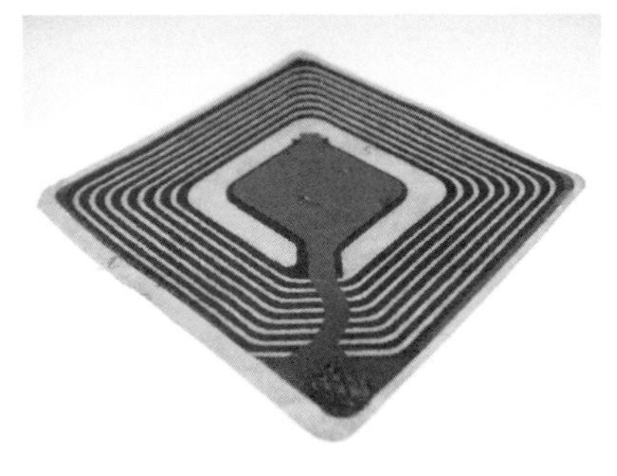

RFID CHIP

Meta Products: Web-enabled product-service networks

The Cloud is a user- and machine-generated stack of data. Like a real cloud, you can't grasp it. Interacting with the Cloud in its most pure form is not really appropriate. We use machines as interfaces in order to read out, understand and build upon this data and we refer to this 'translated' data as meta data or simply put: information. With the advent of WiFi as a wireless communication system, we are able to interact in numerous ways with the Cloud via many devices. However daunting it may seem to some of us, it is technically just a small step from smartphones to any other device. Consumer electronics, industry products, transportation vehicles; they can all be new touchpoints of the web, entry points for communicating with the web, with or without a human operator.

"Companies and organizations that find a way to manage the data deluge and help us navigate complexity will win our loyalty. There is already a huge shift from physical ownership to the sharing of virtual services. Sharing, Co-Creation and Cloud Culture is being driven by a generation used to free downloads and open source approaches. They want to be involved in the process of developing products and services - they demand dialogue!" — ANNE LISE KJAER

iPOD & iTUNES AS A META PRODUCT

Roughly speaking, Meta Products consist of physical elements and web elements. The physical elements can have a sensorial function serving as input for the web elements, after which the web elements perform a certain action, such as storage or editing. The physical elements can also have an actuator function, initiated by the web element, such as informing or alerting the user. The Apple iPod was one of the first Meta Products. Let's take a look at how the meta parts are built up here. The iPod itself serves as an interface, enabling the user to play and browse his or her music library. Nothing very revolutionary going on here you might say, but it got more interesting the moment Apple introduced the iTunes Store as the portal to unlimited music from the web. There you have it, the iPod as Meta Product, consisting of a product and iTunes with an integrated web store, offering you the music of your choice. The added web layer shifted the value to the content part of the product, rather than the object itself. The object becomes merely a means to interact with the content/informational part. So designers will have to rethink many of their preconceived ideas, particularly in the sense of paying special attention to the interaction with different sorts of information and content functions using the same device.

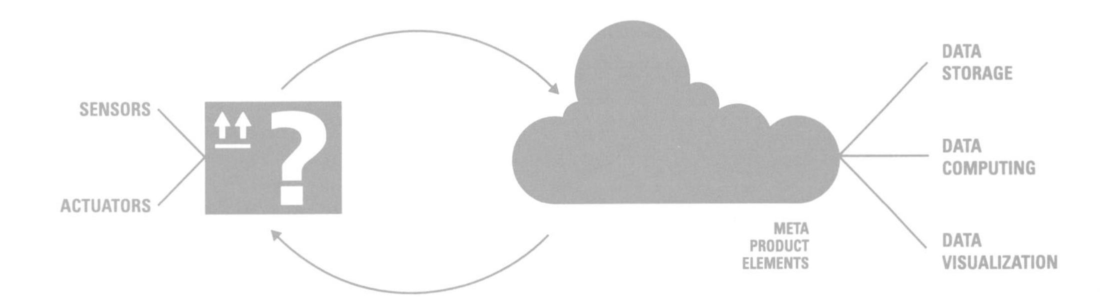

Communication streams

A Meta Product may have different communication streams depending on the goals of the communication in the network (e.g. to give you advice on your food intake patterns or to alert you when your grandmother hasn't taken her medicine), and the touchpoints (sensors, web applications, RFID tags, machines and so on) that are best suited to fulfil those goals. To give you an idea of what these communication streams might look like, we categorized them into three basic types: 1. between a touchpoint and the actual world information (e.g. sensing your body's information, the energy in your house, the growth of a plant), 2. between touchpoint and touchpoint (e.g. machine to machine, sensors to actuators), 3. between a touchpoint and the meta data or information (e.g. smartphone connecting to the API of a web service).

3
COMMUNICATION
STREAMS

① BETWEEN TOUCHPOINT & ACTUAL WORLD INFORMATION

② BETWEEN TOUCHPOINT & TOUCHPOINT

③ BETWEEN TOUCHPOINT & META DATA

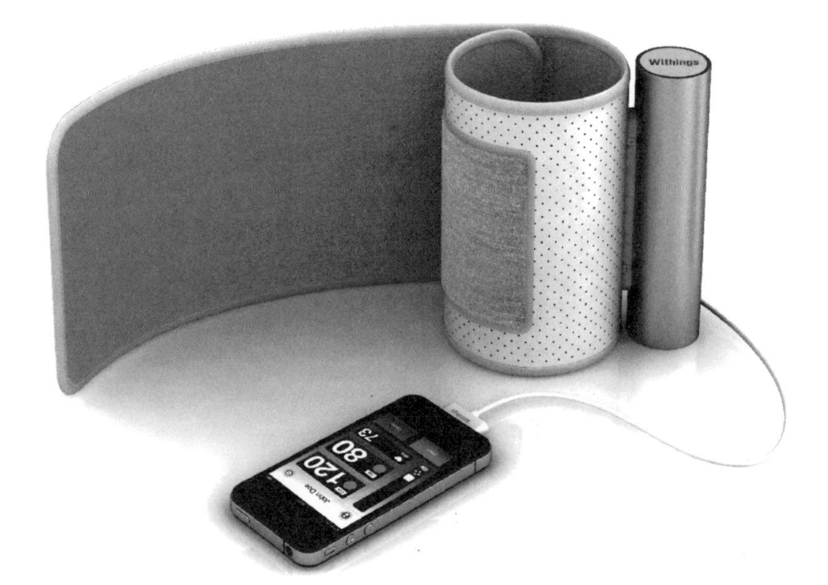

Example 1
With many Meta Products, the information flows from a physical, web-connected product to an online interface. This is the case with the Withings Blood Pressure Monitor, where the combination of the armband itself and the iPhone serves as the physical part, and the resulting digital statistics are presented in an online interface. With these types of products, the physical part often functions as a sensorial input, whereas the online interface plots the results of this sensorial input.

Example 2
Then there is a different group where the information flows in the opposite direction, from online interface to physical product. Take the Olinda radio from Berg London for example, which sends the musical taste of your online social network to a physical radio as streamed music. Now suddenly the online part serves as the input, whereas the product functions as an actuator — playing songs.

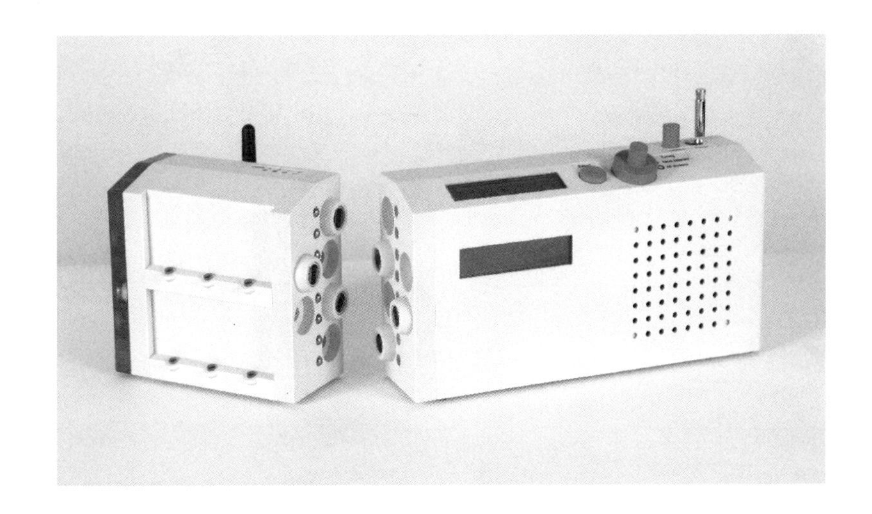

Example 3
Examples 1 and 2 can also be combined, resulting in an information stream that flows from a physical product via an online server all the way to another physical product. Many devices work this way, where the online server merely functions as a forwarding system. One of the oldest examples is the telephone, where a voice signal is forwarded by a telephone company to the other end of the line. Devices using the web as their communication medium can have a far richer information stream than just voice signals. Karotz, formerly known as Nabaztag, can operate this way, originally intented for overseas lovers to communicate with each other through these rabbit robots.

Example 4
Finally, at the borders of meta land there is a group of products that have no direct information flow between physical product and online interface. However, some sort of connection is made between these parts. An example is Webkinz, where a consumer gets a secret code with every pet he or she buys, enabling him or her to unlock a digital pet in the online part. You could question whether this is really meta, since there is no information flow whatsoever. What's for sure though is that the online part adds to the total product experience, making you reconsider what the boundaries of a consumer product are.

Levels of automation

Some years ago people (users) were considered to be passive entities, and much research was put into making systems as autonomous as possible. Today, designers consider people's actions as part of the network they are designing. In fact, in some Meta Products, individuals themselves fashion their own service as they use it. This is not a rule of course. In some cases, the network might require such high levels of efficiency and quality control that it would be better to leave it automated. Magerkurth[3] speaks of system-oriented technologies, which are applications that have the power to 'reason' and exhibit automatic behaviour on the one hand, and people-oriented technologies, in which the applications are dedicated user profiling services (awareness, notification and user interface services for integrating the human in the service), on the other.

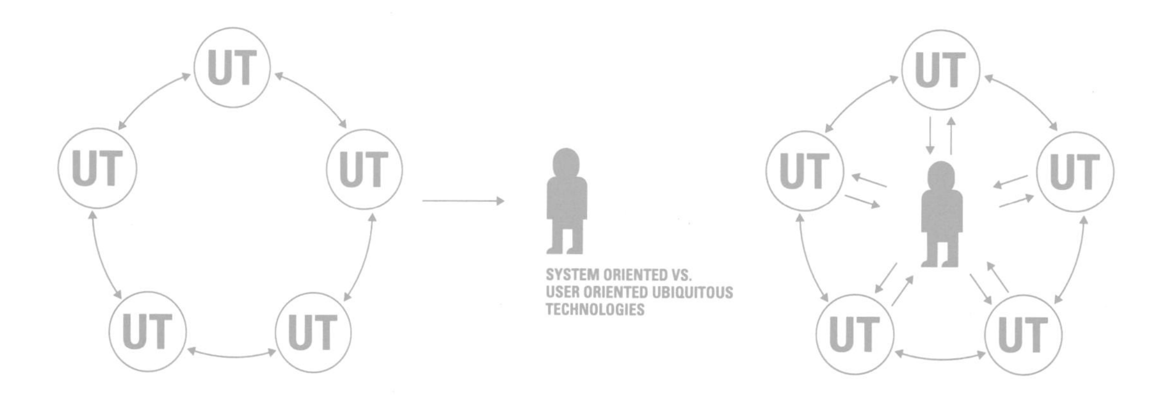

SYSTEM ORIENTED VS.
USER ORIENTED UBIQUITOUS
TECHNOLOGIES

1.4 SOME APPLICATIONS

Sensing wearables

This field consists of wearable devices that measure characteristics of the human body and/or the environment by means of sensors. The accompanying web element often consists of data visualization and analysis of what has been measured.

Ambient intelligence

Intelligent systems that operate in the background (hence the term ambient) within an environment without people noticing it. These systems often consist of multiple sensors and actuators communicating with each other.

Ubimedia

The world of Ubimedia is all about tagged objects, mostly by means of RFID tags, AR tags or QR codes. Readers must be used to retrieving the information within the tags. These readers can be mobile phones, RFID readers, etc. The information that's being retrieved is mostly URLs, linking to rich internet content, mainly for entertaining purposes. The great thing about Ubimedia is that almost any existing product can be enriched with an information shadow, adding a layer of information that was not possible before. In this way products that up till now have just had ornamental or aesthetic value, now get extra value as a result of this online domain.

Identification

Retrieving the identity of an object (for example, tracking and tracing packages in transit) is increasingly being done with the help of web databases. Typically through RFID tags, an object is scanned and matched with an online database to retrieve its identity. A spin-off group of this field consists of so-called 'digital business cards' such as Poken, which are able to swap contact information and identities of social accounts between persons.

POKEN

Navigation and location

Navigating while on the road involves GPS and physical interfaces for pinpointing your location. Dynamic information such as traffic info is downloaded directly from the web, which makes these systems true Meta Products.

Monitoring and tracking

Monitoring and tracking systems often consist of a combination of sensors that measure the condition or state of certain objects. The results are uploaded and plotted in a web interface. These are much like the sensing wearables, with the difference that these systems focus not so much on the human body, but more on objects and their environments.

Games

An ever-expanding group in meta land is the one of toys and games. Combining physical products with the web in whatever kind of way will likely open up great opportunities for optimizing the experiential element of play.

You might now be wondering how you can design these information streams and the levels of automation of your Meta Product successfully. Well, we suggest that you start using a reflective practice to identify the aspirations and the motivation of the people that will use the network, avoiding as much as possible a technology-push perspective. In chapter 5 you will make acquaintance with a method that will guide you through this reflective approach. In the following chapters you will learn more about what's going to be possible in the future, the challenges you will encounter as a designer, and our ideals behind designing for our connected world.

CHAPTER 1 — SUMMARY INSIGHTS

1. Change is the end point of fulfilling aspirations and the starting point for building new ones. How exactly these aspirations are created in our minds and in the fibres of our society is a rich recipe with many ingredients.

2. We see how greater sections of our society with access to so much information today are looking for truly better alternatives in everything they do. They're looking for fair options, customized products and relevant information that help them to achieve their individual aspirations.

3. We are building new aspirations that demand increasingly clever ways to use the potentially endless amounts of information we can track, sense, measure, share and produce via the constantly improving ubiquitous technologies. This will result in breaking down many paradigms of ownership, value, relevancy, temporality, globalization and social structures.

4. The transition period wev are living in is bringing information-fuelled products and services that will be around us as a network whenever we need them.

5. We are making the shift to a society that is no longer determined by the material multiplication of industrialization (product), but by the information generated by our actions (where a product — in the industrialization sense — may not necessarily be involved).

6. Meta Products are dedicated networks of services, products, people and environment fed by the information flows made possible by the web and other ubiquitous technologies: web-enabled product-service networks.

WITHINGS BODY SCALE

Withings is a French company founded in 2008 that specializes in bringing new functions to everyday objects through a connection to the internet. One of the products they sell is a WiFi-connected body scale. The scale can be used by up to eight persons, each recognized automatically by the scale. Every time you weigh yourself, the body scale measures your weight, lean and fat mass automatically. On the product itself, you see your weight data instantly. At the same time, your measured data is wirelessly transmitted to the Cloud.

It becomes more interesting when we look at the accompanying web applications that read out and display your information in different ways. A graphical interface presents the evolution of your fitness over time. The fact that you can see the differences of your weight over time gives you more insight into your behaviour patterns. The application is also used to determine your personal fitness programme and to monitor if you meet the goals that you've set.

For being just a product on the bathroom floor, Withings has succeeded in adding a whole network of services to it by connecting it to the web. For example, Withings has partnered with other lifestyle and health services that offer to help people achieve their weight management goals. So, when you use the scale, you are actually powering the network of your favourite services, food providers and other people with the same goals. The Withings body scale is a nice example of an everyday product that becomes enhanced by connecting it to the Cloud. It's not merely a gimmick, Withings has clearly thought of the value that is perceived by this new interaction. Simply adopting a different way of looking at your information can be of huge value.

That makes you think: what other existing, "normal" products could benefit from a WiFi connection and an online service that goes with it? Which connections and data streams could be useful? Could something be gained by just visualizing the existing data in a different way?

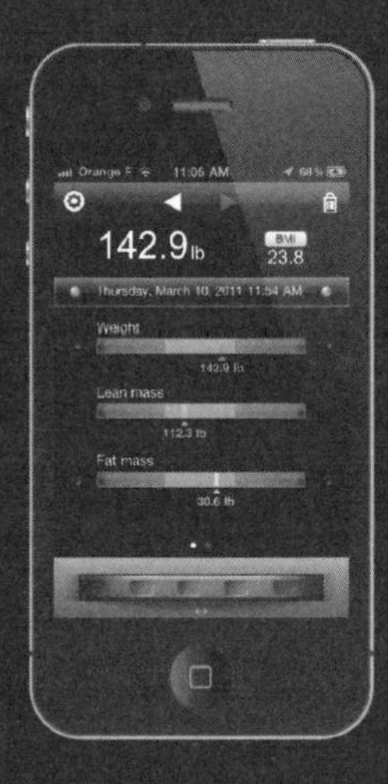

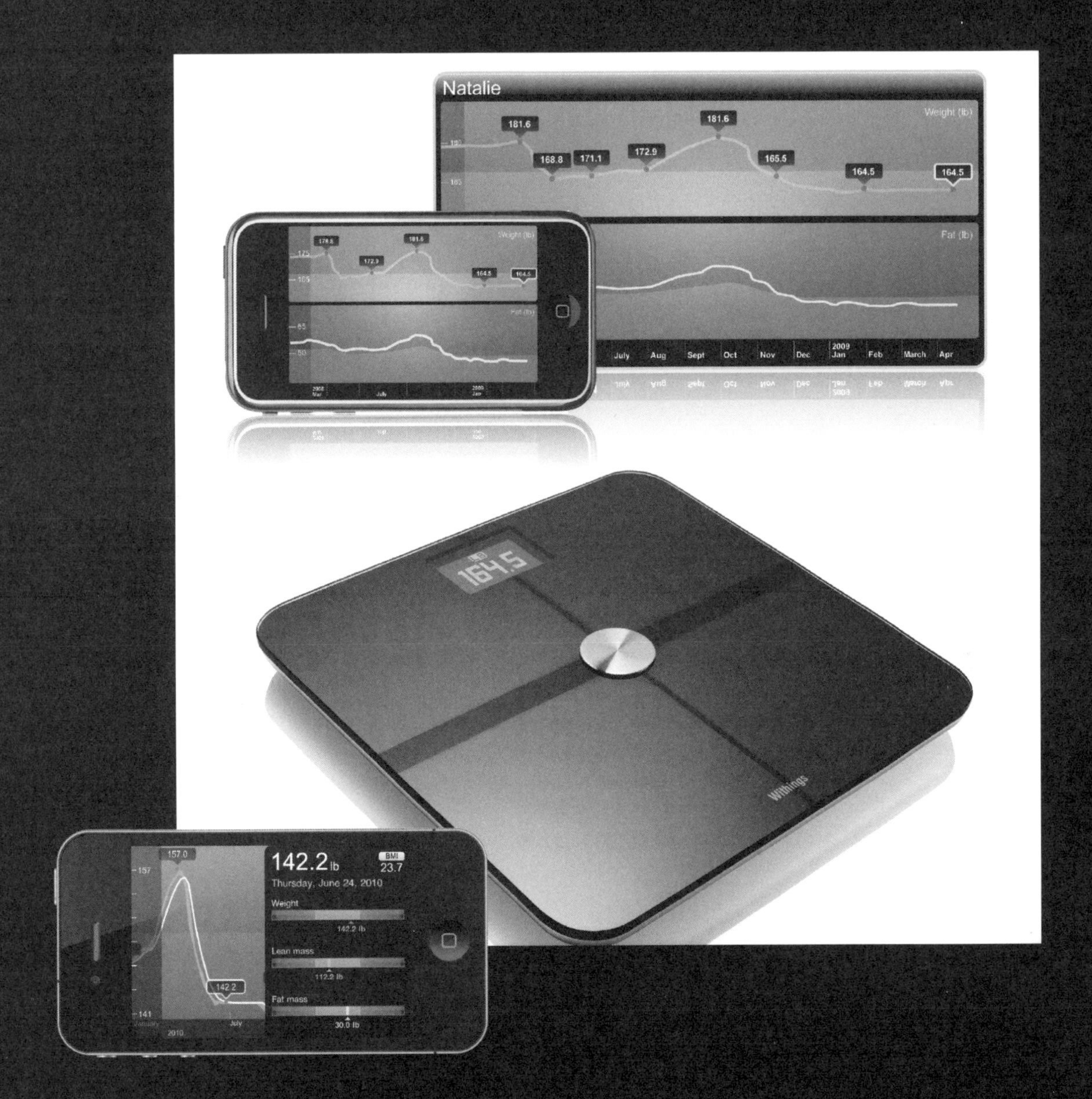

IMAGES PROVIDED COURTESY OF WITHINGS SAS.

NIKE+ PLATFORM

Nike+ is a platform for runners that measures your running statistics and compares your results to your peers. The product launched back in 2006, when it consisted of a small sensor and an iPod kit. You would put the sensor in your shoe, and the iPod kit would read out the data transmitted from the sensor. Since then, the focus has shifted away from the sensor and more towards the platform and making available more options to collect data. For example, the sensor is no longer needed if you have an iPhone with built-in GPS and accelerometer. Via a motion sensing engine developed by MotionX, the iPhone alone suffices to collect the required running data.

Nike+ has been one of the pioneers of Meta Products. Being the shoe giant it is, it launched its running platform to the masses in a relatively short amount of time. The notion that lots of people at the time already had an iPod, and that doing exercises was becoming a trend, was a perfect match for the Nike+ products. They came up with a clever way to connect an existing product, to a complete internet platform. By just placing a small sensor in your shoe, it became connected and transformed your shoe into a monitoring device.

The interesting aspect about the online platform is that it actually encourages you to run outside more often by enabling a network around you. You're not running alone anymore. You can organize challenges online, run against people from around the world, compliment someone for his or her achievements, and so on. All this makes it a valuable network that motivates you to run more (and eventually perhaps become a Nike brand advocate).

Besides selling more sensors and apps for its service, Nike has accumulated a huge database of people who are interested in running. This provides a great opportunity to deploy targeted advertising or come up with new products and services. Nike+ platform is interesting in the sense that it's a self-sustaining service for both customers and the company.

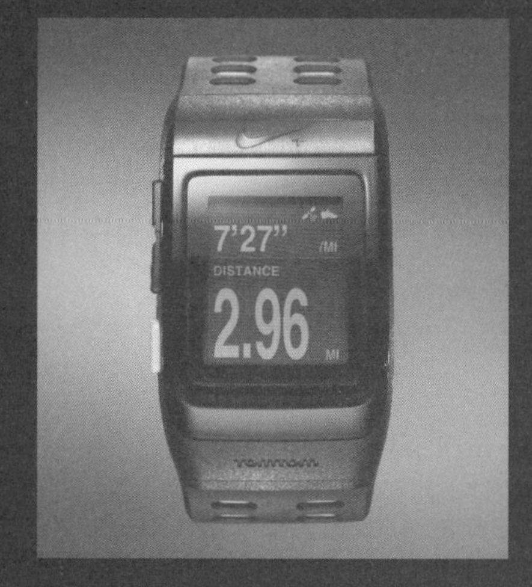

IMAGES PROVIDED COURTESY OF NIKE, INC.

What would you say the future internet and communication technologies will bring?

FEDERICO CASALEGNO
Massachusetts Institute of Technology

The Internet of Things technologies are bringing us "back to human" in a way, back to our local environment. They reinforce the idea of proximity which is not only machine to machine, but also proximity between humans, their local environment and local information. We are seeing this notion of proximity with locative media, near field communication and so on, bringing people back to the local human dimension. We will remain a global village but now with a strong notion with geo-located media, not remotely. This is not only a technological achievement, it is the fact that humans need to be connected with something that they can understand and is close to them so they can touch and manipulate it.

What kind of impact do you think this has on design education and practice?

FRIDO SMULDERS
Delft University of Technology

These new technologies are having an impact on design practice already, and it could have more impact on design education in the sense that we maybe need more courses that make students literate on particular elements of this topic. However, I am not sure this will have an impact on the fundamentals of design education. Designing still remains designing. Of course you have to be updated with the current tools and developments and get additional knowledge. Maybe design practices will be changed to a far larger extent because it is not about products anymore, it is designing intangibles and it means that if we prepare students for the design practice we might have to update our literature, because now design literature about designing intangibles is pretty weak.

FRIDO SMULDERS
Delft University of Technology

New technologies can turn potentially everything into data, but the key is of course to interpret this data. So this is what will put pressure on how the world works, we will have to build data-squeezing skills and get smart interpretations of the sea of data. The fact that we get information in real time allows people to react faster to the information, and to see how others react. If the information is somewhat sensitive for a particular group of people, this group now has the platform to react and organize themselves in a faster way than it used to do. Internet will promote even more cohesion between smaller groups of people.

JEROEN VAN GEEL
Fabrique

It will bring a mobile generation, with Augmented Reality, being able to put an extra layer of interaction over everything around us. Next to that, it will bring multitouch personal informatics, we find it ever more important to get personal data. The touchpoints (points of interaction) are changing per client. All we have to do is keep an open mind and recognize the touchpoints and after that see how things come together. The web experience is just blending in our daily lives.

ANNE LISE KJAER
Kjaer Global

Cloud computing has and is transforming our digital reality. As data and applications move from the desktop to the Cloud, information becomes accessible anywhere, enabling collaboration with distributed teams. Engaging with people via crowdsourcing, social networks and smart apps is now a top priority for senior marketers worldwide. Personal customization of everything is an expected standard. We are always on and ready to participate. From music, fashion and search results to blogs — we are not just observers, but contributors and producers as well.

LUCY KIMBELL
University of Oxford

The impacts on design education include: enabling students to learn how to understand and digest research in several fields including design, management, social sciences and IT; working in cross-disciplinary, cross-organization teams; working as leaders and facilitators; learning how to use designers' tendencies to generate novel ideas and novel research methods effectively within teams and projects; and generally being a lot more modest about what design is or can be, and more attuned to the consequences of designers' decisions.

ALEXANDRA DESCHAMPS-SONSINO
Designswarm

Top-down design just doesn't work. Instead, designers will lead people to understand the possibilities of the technology for themselves. And it can be as simple as tracking your cat, or measuring what you eat because you're on a diet. The challenge for designers today is to make the people use the technology for their very unique needs. I think there is a big confusion about the design role, yes, everybody can put duct tape on the door knob if it breaks, but that's not the kind of design role we are talking about, that's just simple problem-solving. The role of the designer is to see further and not only see a unique personal need but also to place himself or herself in a certain context and decide which tools are the best for the people to create their own projects.

CHAPTER 2 — *Visualize the Future*

aspirational cycle / data / information / knowledge creation / real-time / smart filters / emergent knowledge / computable information / computational objects / thought controlled interfaces / artificial intelligence / personalized filters / sharing behaviour / co-creation / prosumers / proximity / publishing / journalism / online traces / value perception / ownership / smart cities / information city / living networks / micro-segmentation / co-creation

2.1 REFLECTIVE PRACTICE

You don't need to have any super powers to be able to get a good picture of what the future has in store for us. You simply have to observe that the future is a process of different events occurring in the right place, at the right time. We all have aspirations about the future that are constantly changing as they are fulfilled. And we do all sorts of things to realize our aspirations as we deal with our present. So, if you formulate the right combinations or analogies, you will probably come up with a close to accurate picture of the future. Of course this isn't a scientific statement, or magic, it's a reflective practice. In this chapter, we invite you to visualize the future with us and, more importantly, to reflect on the future scenarios that are relevant for your practice. In the previous chapter, we wrote about what we, as humanity, have done in order to fulfil our aspirations up till the present. Now we will present a compilation of major ideas of the future around Meta Products, under the headings: information is the fuel, our sharing culture, and where's the value?

We don't believe it's interesting or useful to tell you about our vision of the future. Our aim is to share our vision of how we can look into the future. And this vision consists of reflecting on the way we experience our past and our present, and understanding how we build our aspirations or ideas about the future. Reflecting in this way might give you interesting forecasts in return.

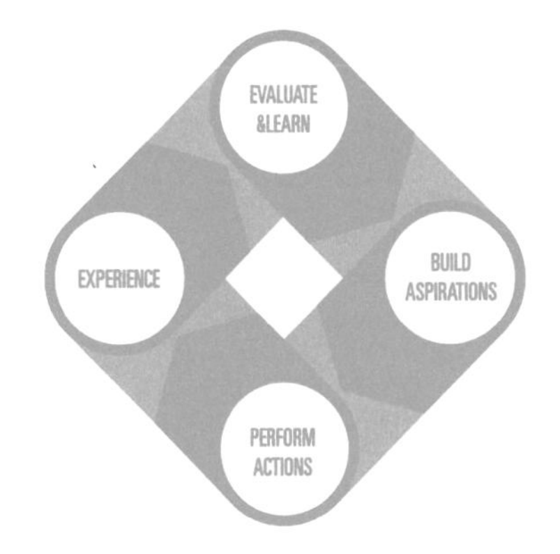

THE ASPIRATIONAL CYCLE

While it's generally accepted that technology changes our lives, the reality is more complex. It's a process, and one that's cyclical. We experience our past and our present in a certain way, causing us to build our aspirations. Subsequently, we perform all sorts of actions in order to convert our aspirations into reality. Through this whole process, we learn, and create new aspirations, which again motivates us to do all sorts of things in order to materialise them.

We've studied the ideas of some of the well-known futurologists and innovators such as Juan Enriquez, Steward Brand and Katherine Fulton[1] among many others, and we can identify a similarity between them that perhaps can explain why they are so good at looking at the future: they are genuinely interested in 'change' and in understanding why we change. They keep abreast of scientific discoveries and research challenges. They are very interested in linking the past to our present and intuitively reflecting on the future. There are characteristics in the attitudes of the great futurologists and innovators that cause them to be constantly dissatisfied with the ordinary, forcing them to look for controversy and confrontation wherever they are. It's not that they are difficult people, it's probably just their way of identifying the real motivators for doing what we do, and why we change. Futurologists and innovators also love serendipity — when you find something you weren't expecting to find, or when you have the ability to link together apparently unrelated facts to create unexpected and valuable new information. Maybe you're a great futurologist and innovator, and maybe you can actually shape the future world through design. So why not try to reflect on some of your ideas of the future right now?

2.2 INFORMATION IS THE FUEL

Data is basically a bunch of symbols that simply are. Data becomes information when you assign meaning to it or when you can derive a relational connection from it. Knowledge is the interaction of different sorts of information with a particular purpose. For example, we would have no temperature if only cold existed. And no darkness without light. So information cannot exist without a context, and without contextual associations we wouldn't be able to make sense of our world or be able to create knowledge. In other words, information forms the basis of our knowledge creation.

In his book 'A Social History of Knowledge', Peter Burke mentions the great impact the advent of printing techniques had on our ability to spread information, and most importantly, he emphasises the interactions that made certain knowledge possible. He provides a thorough analysis and gives many examples of this, including the interaction between scholars and craftsmen in the time of Renaissance Italy: "In the early fifteenth century the humanist Leon Battista Alberti had frequent conversations with the sculptor Donatello and the engineer Filippo Brunelleschi. Without these conversations between such experts it would have been difficult for him to write his treatises on painting and architecture". So, knowledge was — and still is — created thanks to the information we have access to and the way we can share it and change it. Emile Durkheim (1858 – 1917) had this idea in mind already and claimed that knowledge was the collective representation of ideas, beliefs and values. So it seems natural to think we have changed the way we create knowledge if we consider the different ways we access, share and change information today. Today, we are witnessing how the web is enabling interactions that weren't possible just a few years ago. The web makes it possible to

ANCIENT DATA

A SOCIAL HISTORY OF KNOWLEDGE, BY PETER BURKE

access, share and change information. Google, for example, links it for you in an instant.

"Cloud computing is transforming our digital reality. As data and applications move from the desktop to 'the Cloud', information becomes accessible anywhere, enabling collaboration with distributed teams. Engaging with people via crowdsourcing, social networks and smart apps is now a top priority for senior marketers worldwide. Personal customization of everything is an expected standard. We are always on and ready to participate. From music, fashion and search results to blogs — we are not just observers, but contributors and producers." — ANNE LISE KJAER

Wolfram Alpha takes a slightly different approach. It's a search engine but doesn't work like Google. It offers real-time computable interpretations of a large knowledge database as answers to any question you enter. Its goal is to bring expert-level knowledge to everybody and contains trillions of elements of data. It achieves this by making all sorts of information quantifiable, so that it can compute solutions on the fly. You can ask it what the weather was on a certain date at a certain location, or how the planets will align in the future. You can also get detailed nutritional information about certain types of food or how many calories you burn when you go for a walk.

WOLFRAMALPHA

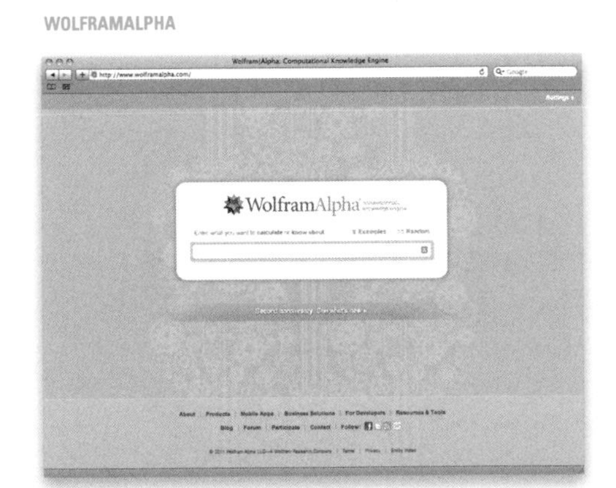

This trend of having more and more information available in a computable way is making it possible for people to build knowledge faster and in ways not previously available. Basically, information about everything is readily available. We could be Aristotle's and da Vinci's Homo Universalis now. We're getting there by building smart ways of filtering and interpreting information that is useful and meaningful to us.

"...the future internet should not only be about quantity but also about reliability and quality of information in any form." — JAN BUIJS

All the information that is being generated in the entire world is giving rise to new forms of knowledge. With the help of Artificial Intelligence (AI) technologies, we can now have access to data about data, which we can also call meta data. What happens is that there's more value in meta data than if you simply added up all the raw data. For example, one of the new forms of stress in our modern society comes from the fact of having to take decisions based on a multiple set of options. We have to choose and therefore we will always be somewhat unhappy about the bunch of other options we had to leave out (a kind of opportunity cost). AI might offer a solution to that: imagine you want to travel around for a year, and you have to make all sorts of decisions such as choosing the destinations, the types of transportation, the budget, and so on. A huge database of travel experiences of real people and AI techniques will filter everything into two or three customized options to suit your best interests pretty accurately. As we've said, it's not simply a question of adding up individual data, the whole is greater than the sum of its parts. This phenomenon is called emergent knowledge. The more traveller experiences there are in the database, the better the service will be. Now add a real-time factor, because not only would you like to get the best travel options but you might like to get them 'on the go'; the service would then be on the Cloud and should be able to react immediately, so that you can be spontaneous and get authentic experiences.

"Immediacy is compelling and addictive as it gives us a sense of living in the now. Why wait if you can shop, bank, chat and check headline news on the go?" — ANNE LISE KJAER

All this holds true for the meta data that is being generated in the Cloud right now with all the things you do when using online services. Maybe the important question here would be the one Lucy Kimbell puts to all of us: "What can these new forms of information do for us and with us? And who is actually 'us'? Placing these questions as a central point is important to recognize the value of it". Additionally, the emergent knowledge we are generating, particularly on the web, is increasingly becoming the most important collective representation of ideas, beliefs and values. If this is happening now 'inside' the web, imagine what will happen when the web comes out and gets embodied in more and more objects and spaces? New associations and information about information will be possible. But how will this happen?

You can observe emergent knowledge in the behaviour of ants. If you look at an individual ant, it's not such an intelligent creature. Basically an ant uses a simple algorithm to find food. It does so by depositing a certain substance in the ground once it has found a food resource. Other ants are attracted to this substance and follow the same trail. Eventually, they end up at the same food location. Looking at the anthill as a whole, you will see all kinds of paths to food sources emerging from the central city of ants. From this helicopter view, it seems as though it is planned or managed by a central identity. But this system simply originates from the vast number of individual ants doing their own thing. The fact that there are so many of them, and that the ant society as a whole has a specific purpose, enables a new emergent behaviour to arise.

ANTHILL

EMERGENT KNOWLEDGE OF ANTS

FOOD

FOOD

FOOD

Information is readily searchable because it can be encoded into a digital form, enabling computers to work with the information and automate search processes. In other words, information is computable. Like information, objects are becoming digital information elements too. This probably started around 2004, when the US government launched a programme to implement RFID chips in their military vehicles. You might not think much of RFID itself; after all, it's just a little chip, it's usually passive, contains an identifier code of an object and that's it. As simple as it may seem though, this chip, and more importantly, the information on it, was the first step in making everyday objects computable.

Building on this technique, you can embed more information than just a single identifier code. Imagine a bottle of milk being transported from a farm to your local store, after which you buy it and put it in your fridge. At each stage of the transportation process, the information on the bottle can be changed. You can for example accumulate all the intermediate locations of this specific bottle on the chip. Your bottle then contains its own history. This type of information is sometimes called the 'digital shadow' of an object[2]. Other objects can read out this information and will then react in a certain predefined way.

DIGITAL SHADOW OF AN OBJECT

The widespread use of techniques like RFID has sparked a huge paradigm shift. When objects are digital entry points, the worlds between objects and the internet merge. Objects will get interfaces which can grab digital data from the Cloud. But what about online data that exists only in the Cloud? Sure, you have your social life on internet, but how social is staring at a flat screen actually? At the moment we're stuck with our screen — the hub between the physical and the digital. We believe this will change, and our hunch tells us that having more computational objects in the future will depend on our aspirations to have immediate services wherever we need them. Unless of course you sincerely believe that spending our lives in front of a screen is something we aspire. We don't think so. Yes, we love our digital services, and yes, we have created our connected world where we can talk to anyone and get all the information we want. Now, we just want all those digital services to come to us in such a way that we can really identify with them and that our lives will benefit from them instead of the other way around. Maybe the fact that the web is spreading out into computational objects is a sign of our basic aspiration for going back to our most natural interactions with the world. Computational objects may be a way to get there.

"The core of the Internet of Things is to seamlessly gather information about objects in the physical world and using the information in multiple applications. Information collection about the origin of goods, location, movements, physical properties, usage record, and context can help enterprises improve business processes, and also create new ones. Needs within a variety of sectors can be addressed, such as remote health monitoring and diagnostics, safe and independent living, intelligent traffic management, improved environmental monitoring, and adaptive energy management." — ANNE LISE KJAER

With the rise of computational objects, for example health monitoring devices, it's now possible to analyze and quantify ourselves. In essence, a blood pressure monitor reads body-generated data. Perhaps that may not seem so interesting, but

EMOTIV
EPOC HEADSET

look at what brain wave monitors potentially can do. The Emotiv EPOC is a headset that can measure your EEG patterns in an unobtrusive way. Emotiv itself calls it not merely a brain wave monitor, but a brain computer interface. The headset makes it possible to compute with brain-generated data. Already it's possible to control a computer game by mere thoughts. Or to type by thinking.

Taking this a step further, we can think of a whole lot of possibilities. Just through directing our thoughts, we can control our computers, our TV, our lights. What about information going in the reverse direction, i.e. fed back into our brain? While neurologists haven't got very far in discovering how our brains work in a detailed way, there's lots of progress being made. For example, the Laboratory of Neural Prosthesis at Illinois Institute of Technology in Chicago is developing an intra-cortical visual prosthesis, which essentially is a wearable bridge device that links hardware camera imagery directly to your visual cortex through the electronic stimulation of neurons. As a thought experiment, imagine that the visual module you're reading now can reliably implant visuals in your brain. Also think of a thought-controlled computer that has internet access to all kinds of imagery. If you link the two, it's suddenly possible to think of a purple cow and you would actually see it in front of you. Much like synaesthesia, in which one sensory pathway is linked to another (hearing colours is an example), you could wire up a brain that has extended functionalities. Ray Kurzweil strongly believes in this blend of humans and machines and calls it human transcending biology. This may sound far-fetched on the face of it, but if you think carefully about the way we're doing things now with the web and our smartphones, maybe we are slowly getting to the stage where we feel comfortable being so close to our technology. So let's take a step back and observe how the more data our technologies can translate into sensible information, the more 'intelligent' these technologies become. Until now we've seen that data will be almost everything and everywhere and, more importantly, easily accessible. But how do we make sense of these massive heaps of data? Which data can become relevant information? It's not feasible to let a mere human interpret it anymore. But even single computers can't keep up with collecting all the data that's being generated. It's simply too much.

HUMAN
TRANSCENDING
BIOLOGY

To make sense of all the data floating around in the Cloud, we might need to interpret it in an automatic way. That's why the future world will heavily rely on AI, because AI makes it possible to construct meaningful chunks of information out of large piles of data. AI can convert raw audio files into written language. AI can translate languages on the fly using linguistic processing techniques. AI makes it possible to automatically detect faces and objects in photos that consist of raw pixel data. And much, much more. AI enables us to get information on a level that makes sense to us humans.

We can go one step further though. What about another level of intelligence on top of this information layer? Now a whole new world of possibilities opens up. Let's consider a simple example: imagine that you use images or videos the same way you use words when Googling something; you'll just take a video or a picture and ask the Cloud about information on it. Additionally, the Cloud will know your location and will make contextual associations that might match your interests. Now we're getting there; all photos today contain some kind of information like the brand of the camera, the shutter speed used, but also geo-information. All these photos are being uploaded and are readily available through services like Flickr. Now imagine a web service that continuously monitors new photos and, by using AI techniques, automatically tags objects in these photos. Now you have a giant database of objects coupled with their geolocation. So, by using countless photos from anonymous people around the world, you can actually derive valuable information about where certain objects are in the world. You could even trace the transportation history of an object this way. AI could interpret data for us on whole new levels and give us information about the world that wasn't possible before. Of course at the moment there are technological issues concerning capacity, reliability and, more importantly, there are unresolved issues regarding privacy and security. These issues are mostly associated with policy making and corporate interests, but we believe you can play your part by designing Meta Products in which the manipulation of information is transparent. Naturally, this is a great challenge, comparable to the kind of challenge faced by product designers needing to take into account sustainability and environmental protection. These types of issues tend to seem so overwhelming that it becomes difficult to understand what kind of influence we can have on them — either positive or negative. However, we believe that if you really co-create with empathy and work in a transdisciplinary team, you will be able to foresee abuse of information, and figure out a way to prevent it. Think, for example, of interactions that people can control and where the options of sharing or not sharing information are clear. Perhaps if you — and others like you — embrace these simple ways of working, you might actually prevent commercial or political interests being imposed on people's privacy and security.

In chapter 4 you will find more about transdisciplinary teams.

"The Cloud is so disruptive in terms of how it's affecting all areas that it's having a real big impact on security and how enterprises manage their data centers today[3]*."*

— *RICH MOGULL*

At the Mobile World congress 2011 in Barcelona, CEO & chairman of Google, Erik Schmidt, said that it is reasonable to expect some sort of artificial intelligence to grow on top of all existing platforms in the next 10 years. "The more Google knows about you, the more personal information you will get." Is this our future interactive process of knowledge? Google and others may see a future in which you will never get lost because you will always have the route information with you, or you will never be bored because there will always be entertainment available, or you will never be stupid because there will always be suitable solutions for you to apply; or they might even say you will never be alone because you will always have contact with your friends or with artificial intelligence that knows you better than anyone else. We could see these scenarios being possible. However, it won't be technology-pushed; the technologies of artificial intelligence will become a part of our lives as a result of the way people build their aspirations and when they get to recognize these technologies as a way to fulfil them. What is unquestionable is that the interactive process of knowledge will be even faster, and that building more, and more meaningful, filters to use the endless streams of information will become absolutely crucial. In chapter 3 you will find more about 'meaning' and how it is related to people's perception of value.

By using AI, we can build smarter products that can at least partly behave autonomously. A good example of this is eCall. This new system — to be launched by the European Commission — is an automatic emergency system for cars. If you are involved in a road accident, the need for hospital care will be immediate. But if you have just been in a collision, you will probably not be in any state to call a hospital yourself. eCall is a module fitted in a car which will be triggered on impact. When triggered, eCall automatically sends information (e.g. GPS location, sensory data, airbag deployment data) to an emergency agency nearby. This system, that is believed will prevent 2500 deadly crashes each year, will be deployed and made mandatory for all cars in the EU by 2015.

2.3 OUR SHARING CULTURE

After the industrial revolution, society in the Western world was long thought of as playing two distinct roles. On the one hand you had companies producing, and on the other hand you had people consuming. Today, however, we are more closely linked to producers than ever before. People are actually voluntarily providing data to companies, and are willing to be public in many new ways. And on top of that, they like to create the services they use and to have control over them.

But why would they do this? Why would you do this? It seems that people have an intrinsic motivation to share. Whether it's emotions, ideas, or just useless messages, people like sharing them. Maybe just like in normal social situations, we like to talk about ourselves and our feelings. It's no different on the web, except that on the web we can share it with virtually anyone anywhere. And so our sharing behaviour has become multifaceted.

"...due to the fact that you can share already so much information with everybody, you may want to share deeper or more meaningful information with a specific group of friends. This is just my feeling of course, that what is happening with internet will promote more cohesion between smaller groups of people." — FRIDO SMULDERS

Maybe this is too brief an explanation to describe the social roots and impact of our digital sharing behaviour. But as a designer it's enough to know that today is the ideal time to explore co-creation and to delve into people's motivation to share. Sharing nowadays is not a neo-hippie movement, the internet is just providing the tools for people to show to the world what they think and do in a very easy way. Now imagine what will happen when more objects are connected to the web, when the environment around you provides all sorts of interactions with the web. You will be able to use your body as the bio-id and all the information you would ever need will be available in whatever context you are. You are already experiencing this with your smartphone, so we're moving in this direction. But we like to think this isn't just a simple matter of fate. There is indeed an apparent snowball of web technologies and derived services raining down on our society, and we have the ability to ensure that these fulfil our aspirations. Particularly if you're a designer, a creator, a strategist, or a marketer... you can decide to do something about our future interactions. If you design transparent services and truly co-create, more people might benefit from our future technology and we might avoid falling back into our old ways where consumers were simply

PROXIMITY AWARENESS

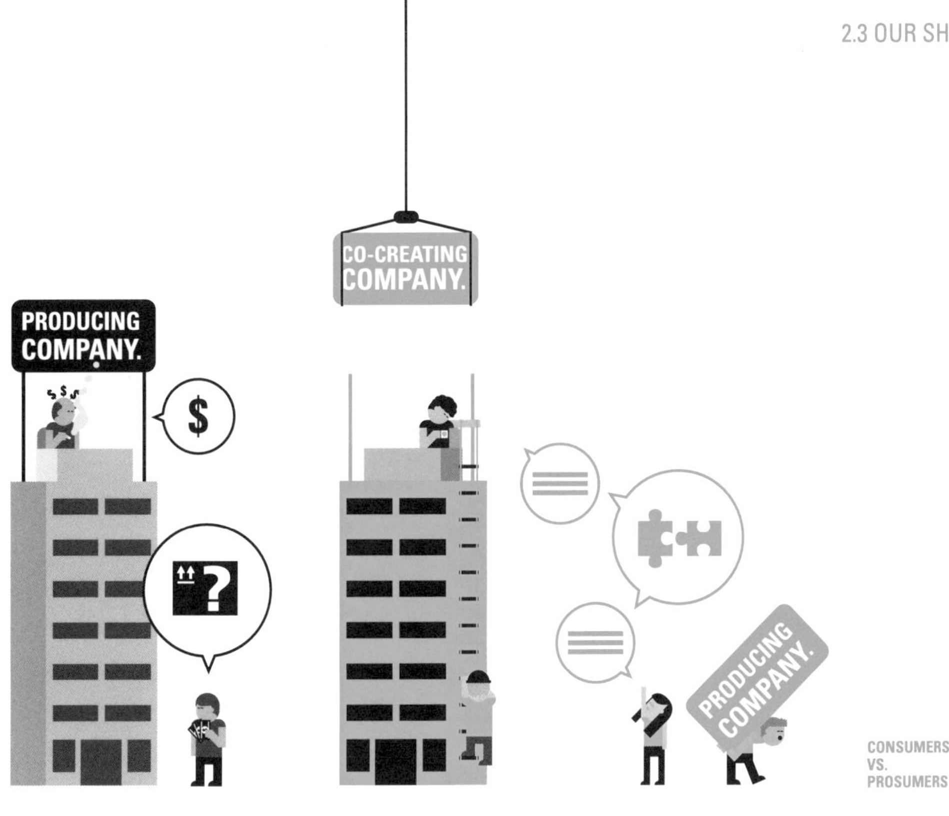

CONSUMERS VS. PROSUMERS

consumers — passive and naive. You might be thinking that not everybody wants to be an active co-creator, and you may be right, but we believe that when people start using Meta Products, they will want to be engaged, to be creative, to express themselves when it's convenient for them.

Moreover, as more computational objects with artificial intelligence are everywhere around us providing us with all the information we need, we will see a change in mindset, from global to local. We were thrilled with being able to know what's happening on the other side of the world in real time. It will still be thrilling, but now we want to go a step further and use this global information for our local needs. And more importantly, we want useful associations that we can identify and have immediate access to. Federico Casalegno, Director of the MIT Mobile Experience Lab puts it this way:

"We will remain a global village, but now with a strong notion of geolocated media... not remotely connected, we were so enthusiastic about that some years ago. Now it's all about geoposition and proximity. And this is not an isolated technological achievement, this is the result of humans needing to be connected with something that they can understand and is close to them, that they can touch and manipulate. This is definitely a new dimension not from the technology stand point but from the human standpoint ...what is happening with the technologies of the Internet of Things is that they will bring us "back to human" in a way... or at least back to our local dimension. The idea of computational objects and machine to machine communication, or the Internet of Things, only exists because we need meaningful proximity between people and our local environment, and get relevant local information." — FEDERICO CASALEGNO

The costs of publishing have decreased dramatically, enabling us to share more readily. Publishers in the traditional sense are becoming almost extinct, because everyone can publish nowadays. And if you only want to publish digitally, it's even totally free. If you stop to think about it, that's really quite amazing. Whereas before you would have needed the financial means, and publishers had to be persuaded to publish your content, you can now publish yourself at zero cost. And this very fact can actually spawn new ideas for traditional publishers: having all these publications readily available in the world makes it quite difficult to find exactly what you're looking for — at least at the moment. Therefore, instead of focussing on publishing itself, publishers would be better off taking on the role of curators. The point here is that roles are changing and new demands for browsing and accessing quality publications are on the increase.

This isn't a matter of business per se, it's a matter of human expression. For example, we can attribute the changes in journalism to the sharing culture and the technological advances of the internet. First of all, journalism is now open not only to photography but to film and web-based real-time interactions and online videos. Today we are seeing new methods of reporting the news, enabling journalists to ask other types of questions and to provide evidence in an active way — whether it involves working with the hi-tech resources of advanced satellite imagery or the crowdsourced imagery and YouTube films from amateurs who are living through wars and tragedies, as well as all kinds of relevant experiences in different parts of the world.

Today it's possible to leave online traces about almost everything we observe and live through. Maybe it would be a good idea if we started to think carefully about these traces when we design Meta Products. What kind of traces, who might use them and why would they? Maybe we won't be able to fully answer these questions, but at least it can open our eyes when we design interactions.

The same mechanism holds true for music labels. For example, Ditto Music lets any music producer publish their own tracks. They handle registration, royalty payments, and all the other things major labels used to do. You can get your music on iTunes and every other online music catalogue by paying a small fee. Everybody can get their music published and make it accessible for music fans at the click of a button. Royalties are paid automatically.

DITTO MUSIC

2.4 WHERE'S THE VALUE?

The value we assign to our relationships or the things we buy, to the activities we perform or whatever we exchange (time, money, knowledge, and so on) varies over time as our aspirations continue to change. For example, over the last ten years Chinese youngsters have increasingly been consuming at KFC, McDonalds, Papa John's and Pizza Hut as a way of experiencing some sort of Western culture. They assign different values to their local food, and even though they are aware of the health issues associated with eating fast food, the value they perceive when going out with friends to eat at Pizza Hut goes beyond any health issue. Additionally, fast food in China is quite expensive compared with some local food vendors. But they are prepared to pay for it, and at the weekends these restaurants are always crowded with Chinese youth eager to live the Western eating experience. They do value their own culture but they are also very curious about all things Western, because these are new and very different from their own. From their point of view, tasting the Western culture is actually quite an 'authentic' experience. Compare this to fast food companies investing a lot of effort in meeting the wishes of Western consumers that are already familiar with fast food and are looking for authentic experiences in healthier and more sustainable foods. We could conclude then that as we achieve our aspirations, we start building new ones and so our perception of value keeps on changing, it is a fuzzy process. Designers should keep a close eye on the fuzzy process of value perception. The best work of Rembrandt (1606 – 1669) was considered by most of his contemporaries to be tasteless and eccentric. This perception prevailed until the 19th century, at which point his artwork was re-evaluated by a new society in which the ideas of Goethe and other contemporaries of the Age of Reason were seeking rational, critical, and genuinely open discussions of ideas. The value perceived in Rembrandt's work changed as the aspirations of people changed.

The truth is that, regardless of geography, today we are looking for 'authentic' experiences. Do you recall when it was exceptional to have WiFi available in a coffee place? Well now it's just a commodity because most coffee places offer it. You just expect to have an internet connection when you're drinking coffee. The fact that the coffee comes in a recyclable cup with a Fair Trade label is what makes the whole thing an authentic experience, and this will probably soon become something we all come to expect when we drink coffee outside our homes. Again, the value perception changed as our aspirations changed.

Today, people's aspirations pose a threat to 'old-fashioned' corporations. Since in fact, these aspirations, as explained in the previous chapter, are the result of our society's quest for alternatives, immediateness, and active participation, and so on. In fact, no single product (in the industrialized sense) can adapt to the fast pace and complex aspirations people are building nowadays. Take a look at the 2011 BrandFinance® Global 500 report, and you'll see that Google tops the list, ahead of Coca-Cola and Walmart. The web, so far, has been one of the best tools for helping you to get and express whatever you want. Your aspirations might push the web to provide interactions that are more natural, more personal and more intelligent. And, of course this will lead to new perceptions of value. What do you think these will become? One of these new perceptions, we believe, is the value of ownership. We mentioned it briefly in chapter 1. What was considered valuable to own yesterday, might be worthless today. Why would you want to own a car at all times, with all the hassle of maintenance, parking space and taxes, when what you really value are the moments on the highway when you feel free and adventurous?

You can see this happening in the games industry, too. Take World of Warcraft (WoW) for example. Computer games used to be products that you would buy and play until you were done with them. With WoW, it's different. It's true that you still pay for the game, but you also need a monthly subscription to keep on playing. This way, WoW delivers gaming as a service. You pay a monthly fee and in exchange you get permission to take part in the virtual world of WoW. The same goes for music. You've probably bought CDs or cassette tapes of your favourite artists. Maybe you've made copies of albums bought by your friends. The point is, they were physical products you had to pay for. With the transition to digital MP3 files, things have changed. Still, the same business model holds true: you pay and get a digital file in return, it's one on one. Spotify on the other hand, has a different business model. It offers you the opportunity to listen to music, not to own music. Its main premise is simple: you pay a monthly fee, you get almost unlimited music in exchange. You can easily see how different this is to the one-on-one deals we're used to.
These new ways are possible because the cost of transmitting digital data is next to nothing. Spotify would never have worked a couple of years ago, because physical albums can not be transported in the same way, let alone fit into your room.

We are witnessing this change because today we attach value to owning moments and living in the 'now' when using our products and services. What would the impact of all this be for you? Our hunch is that the more we value our personal moments, the more our technology will have to be instant and intuitive, and the more 'alive' the network of products and services will be. Maybe there will come a point where everything around us will become part of a living network.

In their essay entitled "Aerotropolis, The Way We'll Live Next"[4] Greg Lindsay and John D. Kasarda visualize the on-going efforts of Cisco to build Songdo in South Korea. An instant, smart, sustainable city and business hub, in other words, a living network: "Picture a Cisco-built digital infrastructure wired to Cisco's TelePresence videoconferencing screens mounted in every home and office, with engineers listening, learning, and releasing new Cisco-branded bandwidth-hungry services in exchange for modest monthly fees. Cisco intends to offer cities as a service, bundling urban necessities (water, power, traffic, telephony) into a single, internet-enabled utility, taking a little extra off the top of every resident's bill. As preached by both Cisco and IBM, the internet will be the next big utility, tying all other utilities together. Hook cities up to the right mix of sensors and software, their thinking goes, and who knows what efficiencies might be revealed? Songdo will be the first city-wide experiment. From the trunk lines running beneath the streets to the filaments branching through every wall and fixture, it promises this city will 'run on information'. Cisco's control room will be New Songdo's brain stem[4]".

AEROTROPOLIS, THE WAY WE'LL LIVE NEXT, BY JOHN D. KASARDA AND GREG LINDSAY

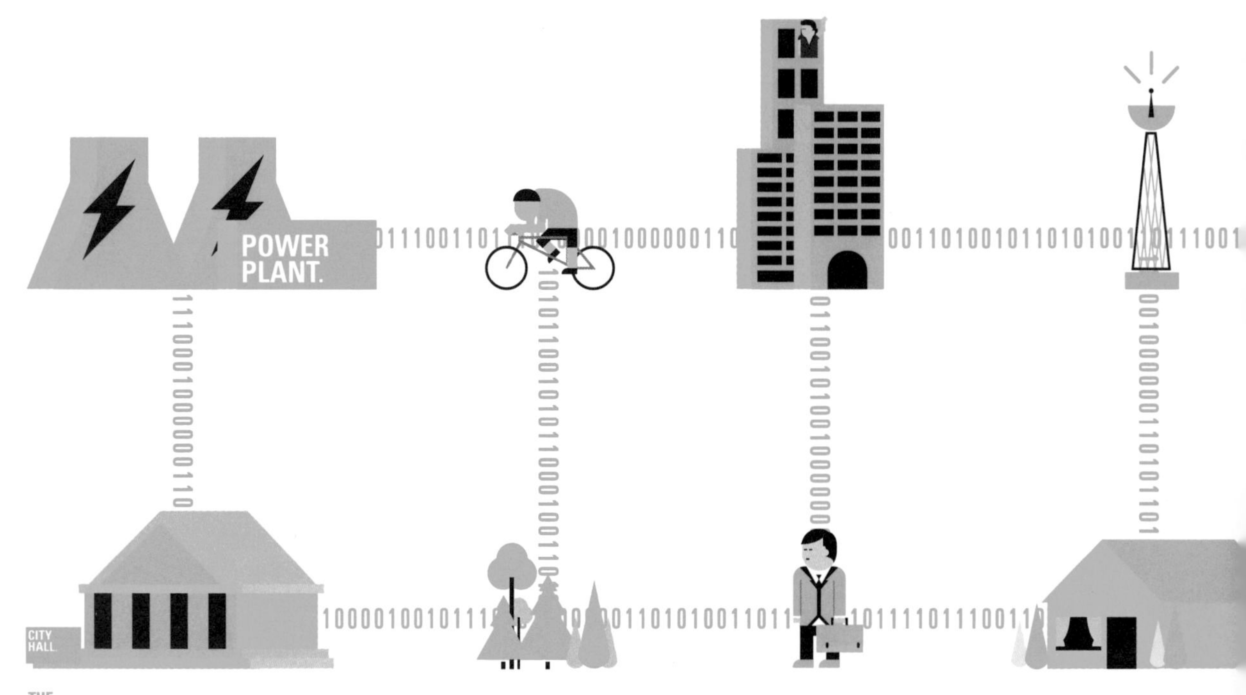

THE
LIVING
NETWORK

This isn't the first time people have tried to make instant cities; in the 50's, the instant city of Brasilia was built based on a different set of ideals, and it failed due to political reasons. Other cities, like Shenzhen and Dubai, were built in record time due to the sudden economic wealth of those countries. Today 50% of the population live in cities and this percentage will probably increase. We're not sure if the concept of an instant city is what we really aspire to in our lives, but the point here is that the city itself will become a living network where immediate and intuitive technologies will come together. And yes, our information will be the 'fuel' for everything. The key for you as a designer is to reflect and observe closely which and whose aspirations are actually being realized.

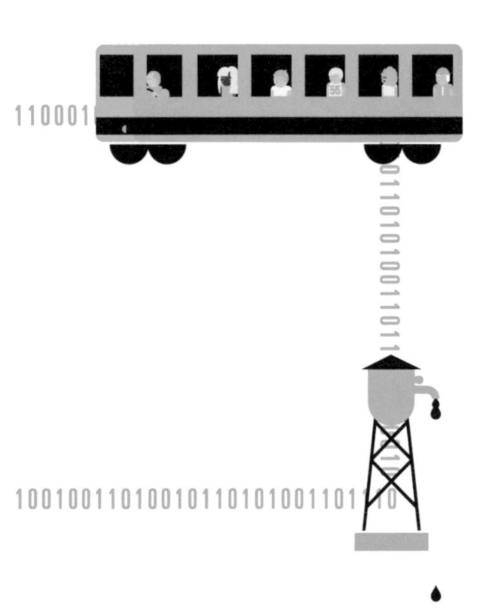

We can see today the potential our technologies offer us to access, share and create information. It is quite evident that the more computable the world becomes, the more connected our world will be. But perhaps the word connected doesn't fully describe what might very likely happen. The way we use our services and products, and actually the way we do many of the most important things in our lives will be nourished by smart manipulations of data performed by the web and other ubiquitous technologies. These manipulations will be the next layer on top of the connectedness we are living nowadays, transforming our services, products and activities into some sort of organisms or living networks where both global and local information (including personal information such as what you're eating or where you are) will stream past, in environments that are always ready for you to interact with. You might want to pay close attention to this because it means that whatever you might design, it will probably become part of a living network of products, services and environments.

"Value is a complex concept that has a multitude of faces, but perhaps in the overall scene of the internet of things there is value in the possibility that it gives us to measure, manage and being told if what we are doing is less efficient than something else. Especially when the effects are on larger scales, such as the consumption of energy, for example. So when this value is recognized it actually changes behaviors... if I can know how if I take two minutes less showering than usual I can actually save resources. Another similar value but in an individual perspective is "personal informatics", the internet of things offers a possibility not only to measure and track my personal information, but to act in intelligent ways. So for example how much I run, how much sugar is in my blood, how much energy I consume... and gives the intelligent tools to make good decisions about ourselves."

— ALEXANDRA DESCHAMPS-SONSINO

Recently, Sigve Brekke, Executive Vice President and Head of Asia of Telenor, mentioned at the Mobile World Congress in Barcelona that technology is enabling us to get so much data from people that it is becoming difficult to distinguish the information that is actually useful. He also mentioned that the future will produce more microsegmentation of markets. According to Wikipedia, "a microsegment is an extremely precise division of a market, typically identified by marketers through advanced technology and techniques, such as data mining, artificial intelligence, and algorithms. These technologies and techniques are used to recognize and predict minute consumer spending and behavioural patterns." We would then suggest adding a layer of co-creation to this microsegmentation technique. By co-creation we mean the active participation of people in the design of their own services and products. Why? Because as we explained in the beginning of this chapter, our value perception changes over time as our aspirations continue to change too. So, it makes sense to strategically put people right in the design process. It doesn't have to be expensive or complicated to co-create, and it might help designers hit the spot right when aspirations are changing. In chapter 5 we recommend some steps and tools for co-creation.

Many ideas exist about what the future holds in store for us. They might provide us with inspiration, but perhaps you want a bit more than that. If so, maybe our recommendations will help you to adopt a reflective practice and envision the future that is relevant for you. It doesn't have to be something that's far removed from you, it could be a client, your neighbour or a group of friends. Reflect on the way they have dealt with their past and how they are building their ideas of the future. Think of how their aspirations change over time; try to co-create with them and to be conscious of the impact of the information that built your interactions can have. We're sure this will help you design meaningful Meta Products with recognizable value.

CHAPTER 2 — SUMMARY INSIGHTS

1. Reflective practice: studying the way people experience their past and their present, and understanding how they build their aspirations will give you interesting forecasts in return.

2. Knowledge was — and still is — created thanks to the information we have access to and the way we can share it and change it. We have changed the way we create knowledge because we changed the way we access, share and change information today. Today, we are witnessing how the web is enabling interactions that weren't possible just a few years ago.

3. Emergent knowledge: all the information that is being generated in the entire world is giving rise to new forms of knowledge.

4. The emergent knowledge we are generating, particularly on the web, is increasingly becoming the most important collective representation of ideas, beliefs and values. When the web comes out and gets embodied in more and more objects and spaces, new associations and information about information will be possible.

5. Information is readily searchable because it can be encoded into a digital form, enabling computers to work with the information and automate search processes. In other words, information is computable. Like information, objects are becoming digital information elements too.

6. Having more computational objects in the future will derive from our aspirations to have immediate services wherever we need them.

7. With the rise of computational objects, for example health monitoring devices, it's now possible to analyze and quantify ourselves.

8. Today is the ideal time to explore co-creation and to delve into people's motivation to share.

9. The use of global information for people's local needs will help designers achieve useful associations that people can identify.

10. Today it's possible to leave online traces about almost everything we observe and live through. As a designer, you can have direct influence on the impact these traces can have.

11. As people achieve their aspirations, they start building new ones and so their perception of value keeps on changing. Keep a close eye on this fuzzy process of value perception.

12. In the future, people will attach more value to owning moments and living in the 'now' when using products and services. The more people value their personal moments, the more our technology will have to be instant and intuitive, and the more 'alive' the network of products and services will be.

13. Cities around the world will become a living network where people's information will be the 'fuel' for everything.

14. Clever manipulations of information will be the next layer on top of the connectedness we are living nowadays, transforming our services, products and activities into some sort of organisms or living networks where both global and local information (including personal information such as what you're eating or where you are) will stream past, in environments that are always ready for you to interact with. You might want to pay close attention to this because it means that whatever you might design, it will probably become part of a living network of products, services and environments.

SIFTEO CUBES

Sifteo is a San Francisco-based company that produces Sifteo cubes. These are modular and programmable blocks that interact with each other. Sifteo's prototypes, called "Siftables", were developed by the MIT and were first shown at the TED Conference in 2009. Since then, the project has evolved into the product that is being sold today.

Sifteo cubes are tiny computers with full-colour displays that sense their motion, sense each other, and can be wirelessly connected to your computer. They are a bit like Lego blocks, in that you can connect the small cubes together and form new structures. Unlike Lego blocks however, Sifteo cubes are aware of how they are connected to each other. Each Sifteo block contains a full colour 128 x 128 pixels display, 3D motion sensing, WiFi, and a 32-bit CPU. To give you an idea of its dimensions, it measures roughly 4 x 4 x 2 cm. It uses a software application as a central hub to facilitate the communication between the cubes. So during game play you need to have your computer turned on, and the cubes must be within 6 metres from it. There are already a couple of games available. For example, puzzle games in which you have to physically rearrange the cubes in order to solve a puzzle. There are also labyrinth games or spelling and maths games.

What's most interesting about the Sifteo cubes is not that they are wirelessly fed with new information. It's that the cubes are aware of their location. Or, to be more precise, their location relative to neighbouring blocks. In a sense, that makes them self-aware. Of course, they can't move around themselves, but as a general notion, it's still interesting.

If we take this a step further, self-awareness is closely related to intelligence. Embodiment, as used in cognitive science, is a position stating that intelligent behaviour emerges from the interplay between brain, body and world. The fact that a product has a certain physical constitution, and is aware of the space it occupies in the world allows us to think about a whole new breed of intelligent products.

IMAGES PROVIDED COURTESY OF SIFTEO INC.

EMOTIV EPOC

The Emotiv EPOC is a headset that measures brain waves. It is a personal (almost intimate) interface, for human computer interaction. The system is made especially for developers, as it comes with a library of code that you can utilize to build your own applications. It is not a bulky machine with all kinds of wires that you have to wrap around your head, but instead it's very compact, affordable and easy to use. Emotiv EPOC won a Red Dot award for its innovative design.

The possibilities of utilizing the headset are endless, it's really a new interface for working with computers. Besides using a keyboard or mouse, you can now control applications by merely using your thoughts. For example, it's possible to reflect facial expressions in your virtual avatar or control the movement of game characters by measuring your brain waves. It's also possible to perform neurofeedback to steer your brain towards a certain state of activity. There have been experiments linking the headset to a car, thereby making it possible to drive a car with your mind.

The significance of this product is manifold. We now have the ability to control by thinking at our disposal, opening up a whole new range of possible interactions. Also, by providing an easy-to-use software development kit (SDK), developers and scientists can now study brainwaves more easily.

By being relatively affordable, the headset is suitable for a large group of people, from indie game developers to universities, or even curious individuals. You could even think of mobile apps that use the headset as an interface. For example, a stress-relieving app that works in combination with the Emotiv headset for a complete monitoring and training solution for personal use.

This Meta Product is a good example of making use of the new possibilities to retrieve data. If you think about it, all the device does is transform your brain signals into a range of bits and subsequently into meaningful information. In other words, it makes your brain quantifiable, it discloses your brain wave data to a format readable by a computer. Having accomplished that, the EPOC headset is an innovative touchpoint for connecting ourselves to the Cloud and other devices. Which products or applications would benefit from controlling by thinking? What would you like to control with your thoughts?

IMAGES PROVIDED COURTESY OF EMOTIV

How would you think our interactions with products and services will be like in the near future?

ANNE LISE KJAER
Kjaer Global

They will be meaningful curated collaborations that empower the individual. Creating a new language to interact with audiences and promoting an innovative environment is key in the 21st century. The concept of consumption stretches normal boundaries. More and more companies explore the abstract ideas of their values and company culture — their cultural legacy or cultural capital. Launching concept products, services and retail spaces and then inviting people to participate in a 'curated' art gallery-style experience is one route to get there.

How would you define the role of designers nowadays?

MARC FONTEIJN
31Volts

Our role as designers has changed compared to 10 or 15 years ago, we are not the experts, the user is the expert. The people that went to a bakery store 100 years ago, they were the experts, they knew what they wanted and they could tell that to the local bakery. It is still the same, except for what has happened is that organizations are now unreachable and they can do whatever they want. So we are pulling the user back to the expert position. We are facilitating and giving them the tools to make them heard and get the important insights out of them.

MADDY JANSE
Philips Research

In the widest definition it means that all the objects in our environment will be connected via the network, in such a way that from my mobile I can talk to my coffee machine and so on. We've enabled all these possibilities with our technology but what will people pick up from this? What can be really sensible? We only have 24 hours a day, from which we work 8 hours, sleep 8 hours, we need to eat too and do many other things. People won't have the time or the energy to be engaged in new interactions if they don't see the real value.

JENNY DE BOER
TNO

Generally, mobile internet will be more integrated in your life: internet coming to you instead of you going to it. We will see that in developing regions a big part of the internet evolution will take place. They will skip the fixed lines and the computers, they cannot afford them any way so they will go for the mobile phones. Therefore, I see that more interesting potential applications and interactions for mobile internet can come up especially in developing countries.

LUCY KIMBELL
University of Oxford

Most future product service systems are likely to have both online and offline interactions, although not always from the point of view of the end user. Designers focusing on the user interface will need to decide whether knowing about the status of networked connectivity matters to the user at a particular point in the service experience.

FEDERICO CASALEGNO
Massachusetts Institute of Technology

Designers today and in the near future will just assume that an interaction with something tangible has also potential connectivity. Embedded electronics into objects is just a reality. Designers have to learn to use this new material: connectivity. Whether to use connectivity into objects or not is just a choice or a design direction to decide. But they should be very careful and very critical towards their decisions by understanding the human dimensions.

FRIDO SMULDERS
Delft University of Technology

The world is changing for the designer. Particularly industrial designers were commonly associated with products and giving shape to things, which is already for quite a while not the case. However our faculty is still pretty much aimed at form and tangible, even though we are already designing systems and intangibles, but we may not call it like that. We might call it "all the other things happening around a product." We will still need to design chairs, tables and pencils, but for some designers Meta Products will be more relevant. For them, collaborative effort will be very important, because the user experience in that case will result from the whole system or network, and not from isolated products.

HARALD DUNNINK
Momkai

Designers have to use their talent of empathy: placing yourself in the situation of the consumer; particularly within interactive design. There are ways to become better at it, and that's by talking to people and always trying to find their perspective. Within our team we try to do it, and talk a lot and we try to see things from the technical side or from the client's side. There are ways to open it up. You have to observe and be interested in what's happening around. To me every other perspective than my own is the best inspiration there is.

CHAPTER 3 — *The Bit and Atom Disruption*

embedded mobile internet / new languages / connectedness / dedicated network / digital and physical connections / dedicated network / engaging interactions / empathy / mental models / social bonds / human-friendly-technology / collective interactions / embedded electronics / enabling relationships / value / design thinking / transdisciplinarity / strategic designers / fuzzy front end / empowerment platforms / business-design thinking / scaling potential

3.1 CHALLENGING DECISIONS

This chapter is all about the challenges you might encounter when making design decisions in our connected world. Bits and atoms have never been so close to each other as they are today. We believe this has opened a window to many design challenges. We're not talking about the choice between bits or atoms, but about a disruption brought about by the relationship between bits and atoms: the moment bits (ubiquitous technology) are embedded into objects and our surroundings, an enormous amount of possible interactions arise. Hence, making design decisions becomes even more challenging.

It's your decisions that define who you are. Your life is a history of decisions, some of which caused you much distress or pain. Lover or friend, give or lend, more or less, feeling or thought, alone or collectively... the simple truth is, we don't really like to make complex decisions. Some of them are so difficult that we occasionally prefer to carry on with our lives pretending they don't exist. But they won't go away until we make a decision. The more options you have, the more difficult it is to choose, particularly when you want to make the right decision that will not only define you but will also affect other people. A doctor's decisions define people's state of health, and a politician's decisions define people's opportunities for education, for example. How do you influence people's lives with your decisions as a designer?

It doesn't matter which type of designer you are, whether you're a product designer, an interaction designer, or a service designer; or whether you take strategic design decisions at your company; you will inevitably have too many options to choose from, and hence too many decisions to take as a result of the bit & atom disruption.

What do we mean by this bit & atom disruption? The more mobile the internet becomes, the more it acquires the ability to become embodied virtually anywhere and in anything. This is resulting in products, ideas, knowledge, services, people, and other technologies talking to each other in new ways. You might wonder whether this disruption really exists, since we can all just adapt to it, just as we adapted to the mobile phone and are adapting to the smartphone. Indeed, it may seem simple on the surface, but actually many things are happening in the background in order to allow new things to become part of our lives; particularly if that new thing is a technological innovation. If you're a designer, you are by nature thinking of designing new things, and if you live in the 21st century, the new things you're designing will potentially involve some disruptive innovation. Why disruptive? Because what is happening with the web will shake up the current way of doing things, at least in many aspects of our lives. Many large companies generally don't react to disruptive innovations directly because these innovations don't represent an attractive market or one that's easy to identify. So what these big corporations are used to doing is forming a smaller organization that can cope with these disruptive innovations more adequately, and bringing their infrastructure and resources into play at a later stage. When people interpret something as being a disruptive innovation, it typically means there are lots of unpredictable things happening. We believe that this is the best time for designers to start turning on their antennas.

In this chapter we will share with you why we think that today there is a disruption carrying new design challenges. We have divided our explanation into four parts: networks, meanings, processes, and platforms. We hope you find it interesting to read, but most importantly we hope it will help you build upon your own reflections and identify your own challenges when designing Meta Products.

3.2 DESIGNING NETWORKS

"Every device will have an internet connection as standard, that's increasingly the case already[1]*."* — SURANGA CHANDRATILLAKE

The utmost mobility of internet will increasingly enable things to connect or communicate with each other, where they couldn't do so before. Things like products, services, technologies, resources, concepts, spaces and more and more people will have new ways of communication enabled by the web and other ubiquitous technologies. What does this have to do with designers? Well, products and services will never be stand-alone anymore. They will be increasingly related to other products and services and to other things, people, and environments via the information they generate. They will be Meta Products. Here you could say that in the fundamental sense, nothing is or has ever been stand-alone, and you'd be right. Everything is related in some way to everything else. A lamp, for instance needs a context to be a lamp. It needs an electrical installation, the physical location on which it can be mounted or hung, but it also needs the objects or the space that require to be illuminated. So indeed, if we look at the world through an analytical lens, everything is part of a system, nothing is really stand-alone. So, are Meta Products actually

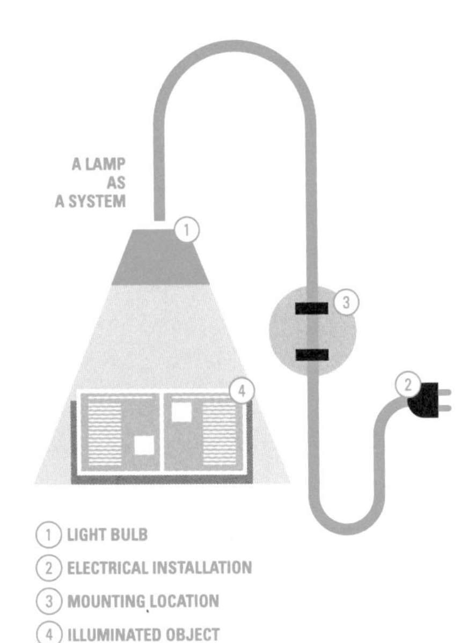

new then? Well, yes they are, because for the first time in the history of humanity, a window has been opened to endless new connections, new ways of communication and to new interactions with things that used to "talk" in a different way. For example, you could say that a chair and a table are not stand-alone entities, and that they need each other, and also "talk" to you in their own language. Just by looking at its shape, you will tell if a chair is comfortable or if you can drag it or lift it, or if it's expensive. Shape, colour and size are some of the typical language elements of a chair. But in the world of Meta Products a chair may also make decisions and talk to the table, to you, to the space around it, and to other objects. Why we would want a chair to do so is another part of the disruption we will address later. But for now it's important to realize how vast the range of new possibilities is. More importantly, some of these possibilities will eventually lead to new ways of building, accessing and sharing knowledge. Which possibilities are we going to take? For whom? And what will the impact of choosing some possibilities be?

You are witnessing the second greatest turning point in the history of humanity since the Industrial Revolution that will exponentially affect, and is already affecting, every single life on the planet. Things will be connected to other things in ways we never thought we would see, and it will happen first of all because it is possible and secondly because we will never stop building aspirations. We will continuously seek to surpass our limitations. What is happening with the web might be one way to help us make many dreams come true.

"All products were already connected to each other, only not by the technology we use now. There was always an intangible system around products. The only thing now is that products can suddenly exchange information and talk to each other, so we can add some sort of intelligence to them. This is just opening more possibilities, if you are a good designer you will be able to identify good possibilities." — FRIDO SMULDERS

If you want to get a better picture of how great the technological advance of the web is, you can maybe try to compare it with what happens in our brain. The brain is still sort of a mystery for scientists, but they suspect that some sort of sudden boost happened in our evolution to make our brains so sophisticated in a relatively short time span compared to other creatures in the world.

What is astonishing is that the brain has the capacity to have a massive number of highly interconnected neurons working in parallel, with an individual neuron receiving input from up to 10,000 others. Creatures that have the ability to adapt to a changing environment — like we humans can

— need a brain which is capable of learning. Our brains use very complex networks of specialized neurons to perform this task.Incorporating a brain metaphor, we could say that in a Meta Product, a group of neurons would be systems (products, services, people, concepts, resources), the points of output and input connection between systems would be the web and the supporting ubiquitous technologies. Neurons and their points of output and input connection define together a network in the brain dedicated to, for example, learning songs; in a Meta Product, the systems and the web define together a network dedicated for example, to shopping, or healthcare, or travelling. This sort of complexity is what designers will be facing in the near future. But this doesn't mean they will stop designing products like buttons, knobs and chairs. It means that designers will also have to deal with the new networks linked to those buttons, knobs or chairs. More than ever before, designers will have to master their non-linear skills of understanding a situation and identifying opportunities along the way. So they should embrace their core specialties and at the same time become the ultimate network designers.

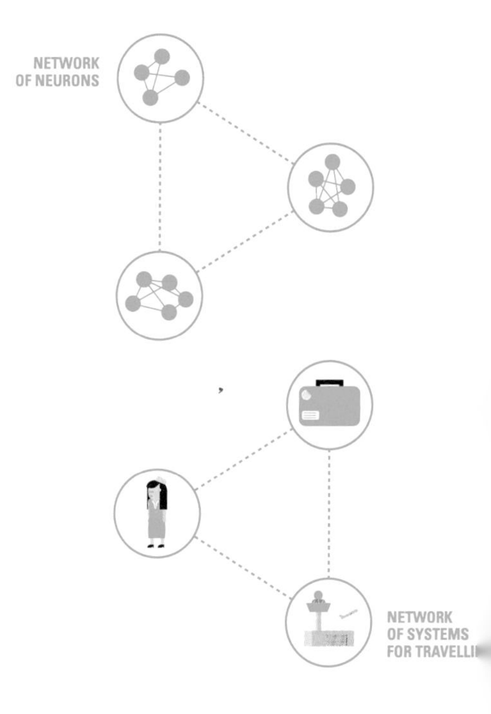

"...no single designer is going to design a complete network, it's more about resources being mobilized, including some specialists who work with bits and some others who work with material stuff. So the challenges are about teamwork, collaboration, networked working, professional competences and knowledge sharing and so on, rather than focusing on the properties of bits or material stuff." — LUCY KIMBELL

Meta Products provide a special focus on the connections that build up a network rather than on the isolated product or the particular service. The connections, or the relationships between information and you, your products, your friends, your business, your client and so on will continue to gain in importance. We can imagine that this might all sound too abstract for you. So, let's make things a bit simpler by breaking down a popular Meta Product. Take the Senseaware developed by FedEx for example. This is a Meta Product that helps the people behind the logistics industry know when a package en route has been opened or exposed to light, its exact location in the transportation process, as well as other information that is visible to different key people in the supply chain to enable them to make better decisions.

"...five years ago, you wouldn't have designed an object knowing that it could be so immediately connected, recognizable, and this object could change and react. Today potentially this happens. So the way of designing is different, you have to consider a new spectrum of variables, also too many variables in which having a very critical mind is necessary." — FEDERICO CASALEGNO

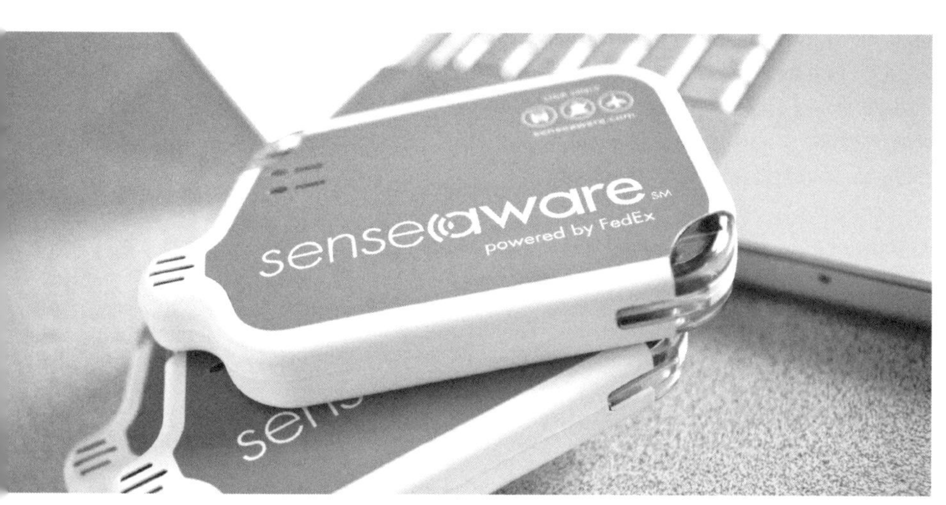

SENSEAWARE

The system consists of a multi-sensored device and a web-based information platform. The device transmits information to automated tools and to the logistics team to help the key people react appropriately. In this Meta Product, elements that are not able to communicate in a traditional logistics network, such as the package en route, now communicate to multiple people in different ways.

If you design Meta Products, you are basically designing networks with connections that switch from digital to physical and vice versa. All this switching may mimic our 'real' relationships, improve them in some way, or create new relationships, and — ultimately — new behaviours. For example, paying online mimics the way we pay with real money, but also creates new paying relationships within countries because the transactions can be done worldwide, all taking place at the computer. In some countries you can pay with your mobile phone, the bank provides you with a unique seven-digit mobile money identifier (MMID), which is given to each account linked to any mobile phone number. You would also need the vendor's number, so money can be transferred to it. Then you would receive a confirmation as soon as the transaction is done. Typical Meta Product scenarios like the one mentioned above will inevitably involve a lot of switching between the digital and the physical. You will be carrying out intelligent measurements, tracking, identifying physical elements and controlling mechanisms, leading to new interactions. You might be thinking that we are already doing all this without any problems. But we believe we need to reflect on the impact of these possibilities both in the design practice and in the way people change their behaviour by performing these interactions.

"I think an interaction is more about what you do and an experience is more about how doing something makes you feel... we let people interact with certain technologies so they can get experiences that help them really see the possibilities."

— JENNY DE BOER

Think of urban planners. They have to constantly think about the collective use of environments. They have to create the proper connections for people to their work places, homes, hospitals, schools, etc. But they don't stop there, they also have to think about social behaviour and cultural identity in relation to the aesthetics of the city. The connections between all those elements foster economic health and sustainability in a city.

We believe that if you start grasping the idea of 'designing networks' you won't lose focus on the relationships that are valuable for people, and this will help you design meaningful Meta Products. Moreover, it will help you avoid getting bogged down in technical issues and instead focus on the value those interactions bring to people.

If you reflect on what people find valuable, it might help you to get both your feet firmly back on the ground. Our friends, our families, our communities and our societies are who we design for. When designing Meta Products you will find that the complexity of a human experience may become even more intricate when the possibilities of connecting and communicating multiply. People are precisely what the next part of the bit and atom disruption is all about.

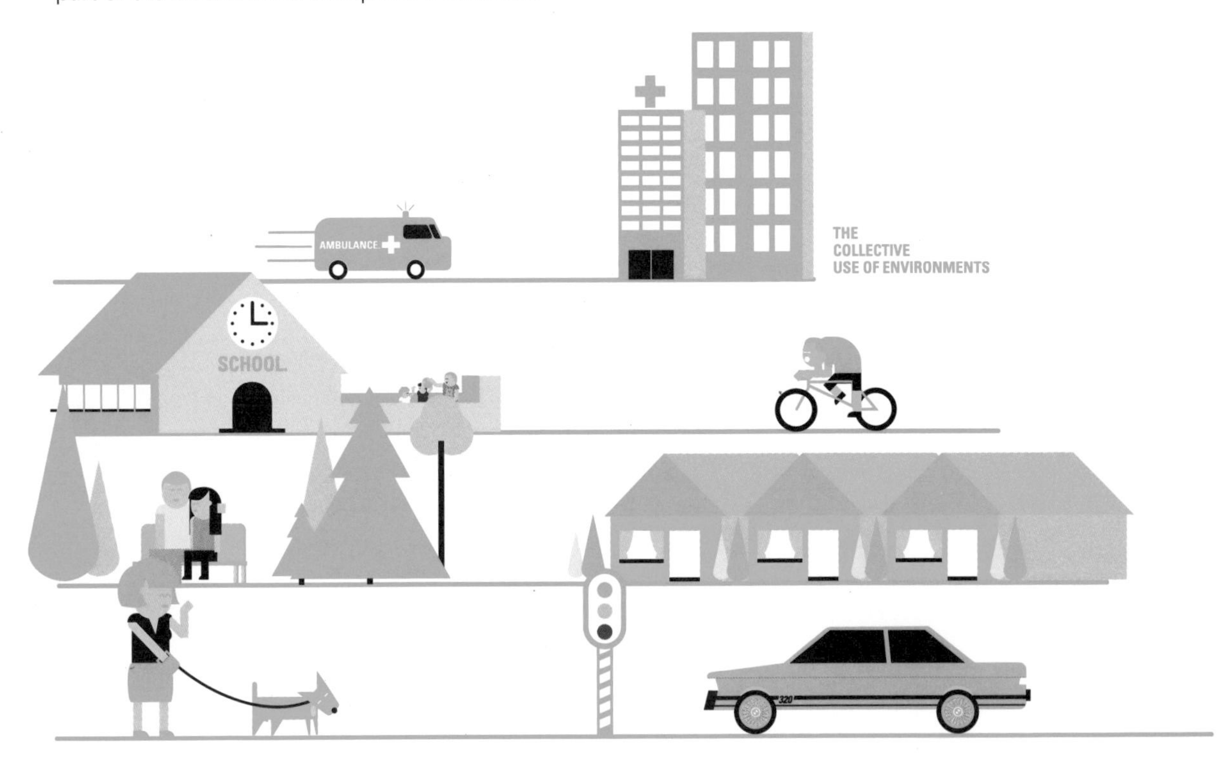

THE
COLLECTIVE
USE OF ENVIRONMENTS

3.3 NEW CONNECTIONS, NEW MEANINGS

In our interconnected world we are seeing increasing numbers of relationships and marriages between individuals from different cultural backgrounds. Today, it is easier than ever to go wherever you want in the world and to communicate remotely. Take two personas, let's say Bo-Bae and Alejandro. Bo-Bae comes from Korea and Alejandro from Spain. They met online and had a virtual relationship for a while until he decided to visit her. Before going there, Alejandro and Bo-Bae communicated via chat and Skype and didn't seem concerned about anything other than meeting each other in person. But, the day came and Alejandro set off on a journey of learning all sorts of new meanings. He just never thought that the things he already knew could have a totally different meaning in another environment. So, he learned to wait until the elderly had started to eat before starting his own meal, otherwise it would have been considered extremely disrespectful. He learned to take his shoes off before entering a house, to not talk back to the police officers, and to bow according to the status of the person he was greeting. Koreans and Spaniards, and so other cultures, have their own language, their own ways of doing things, and sometimes they assign different meanings to the same things around them. All this can produce difficulties in understanding each other. The same can happen with Meta Products.

Things that couldn't communicate before, now can. A table and a chair used to speak a different language to talk to you or to each other before they were able to talk through wireless sensors. Or your measuring scale used to just tell you your weight but not to provide you with an entire customized weight-loss programme, for instance. All these new languages and new interactions, such as the ones Alejandro experienced in Korea, have to be refined until the meaning is clear for the people engaged in these new ways of communicating.

"It's always about 'relevance'. An experience can be very relevant at the moment it's happening, regardless of the technology used. Maybe it's not necessary to have all these technical gadgets at all. Nothing is really missing in that sense because technology is not really the point; it's about what you want to communicate or accomplish now." — SEBASTIAN KERSTEN

Think of a Meta Product as a conversation between physical things, information, people and spaces (and whatever else you can imagine) enabled by the web and other supporting technologies. In a simple conversation there will be a sender, a message and a receiver, and an exchange of information takes place in both directions. So, good communication depends not only on the sender but also on the receiver, it is a relational process of course. Now imagine there are multiple senders and receivers, imagine communities, imagine a network of communities with parallel levels of communication. Imagine on top of that there are all these new languages! Basically, what you have to do is use your most sophisticated empathetic skills as a designer to try to understand the relational process that takes place when people assign meaning to new interactions. You also have to be very alert to what sociologists, anthropologists, neuro-scientists, psychologists and all sorts of professionals that study people have to say.

"It is a delusion that any designer can convey the "right" meaning or design for an experience to play out a particular way. What we know from studies of media and other kinds of consumption is that people bring their own desires, identities and purposes to their activities which no designer can foretell, however much research they think they have done. What designers can do is assemble things into temporal and spatial arrangements with an understanding of how things might perform — choreography and curating are useful concepts here — but with the modesty to understand that people bring their own meaning-making to their encounters with such assemblages in practice."

— LUCY KIMBELL

In 1960, Hollywood producer Mike Todd Jr. released the film Scent of Mystery, which featured a special technology that added aromas to the scenes. It was called Smell-O-Vision, or Sensorama. Films couldn't communicate smells, so they thought of adding this to make the films more attractive to the audience. Soon after the release, the whole concept failed miserably, not only once but twice. In the first trial, it failed because the human sense of smell can get easily overloaded. The air had to be completely cleared of a particular aroma before another aroma could be released. Otherwise, the scents would all blend together into an indistinguishable olfactory melange. In the second trial, the developers of Smell-O-Vision figured out a way to pump out each smell after a new one appeared, but it still failed. This was probably due to the complexity with which humans perceive their environments, where smells are naturally part of the context and not isolated elements linked to cues shown on a screen. The communication failed totally and the experience was just meaningless, regardless of how much their creators were convinced it would be a success. Without really trying to understand how people assign their own meanings to the things they experience, the experiment was doomed to failure.

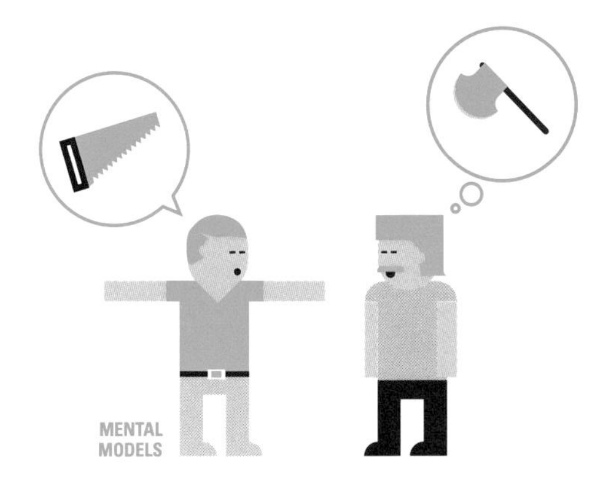

Understanding the basics of meaning

Perception is the beginning of a human experience which may ultimately be meaningful to a person. In linguistics, 'meaning' is defined as associative links in our minds between words and objects and experiences, that result in the formation of concepts[2]. Experiences are "intangible processes of interaction between people and the world that exist in humans' minds and are triggered by new interactions"[3]. Both definitions refer to the human mind as the place in which meaningful experiences exist.

Everything around us is translated into our own personal concepts or cognitive models. According to Norman[4], mental (or cognitive) models are "the internal representations that humans develop of themselves and the objects they interact with in the world." Building mental models is an important component in adapting to the world. Hence trying to understand mental models may help you to design interactions that are meaningful for people. It is also important to know a little bit about how mental models are created in the human mind. Human cognition is a natural information processing system that has many 'tools' with which it assigns meaning to the world, and behaves according to those meanings. But most importantly, all human processes change or evolve during time, even when the interactions remain the same. Because people learn and create different associations with the same things they have been interacting with, as a designer you have to be aware that the relationships in a network are perishable.

Blythe et al wrote in 2003: "It is essential to build design practices on wider analyzes of human nature, activities, and mentality. How such a holistic analysis can be realized is one of the major challenges for modern interaction-oriented, cognitive and information systems science."[5] And it is still valid today.

"Connecting with people requires empathy and vision. It is about the ability to inspire, educate and empower the end user for the purpose of enriching their quality of life. Understanding that people — whether they work for you, use your services or buy your products — have higher standards and more complex decision-making processes than ever before. They also expect more meaningful choices, so whatever your offer: present it an ethical and meaningful package." — ANNE LISE KJAER

The collective magic

Meta Products take the challenges of understanding human nature a step further. The technology of Meta Products is embracing the primeval need people have to connect to each other and to other things. Funnily enough, we do not know much about this need or the impact of it. Apparently, scientists have been more focused on the individual's behaviour, brain, psyche and emotions, but a few have recently started to explore the collective or social human nature as an entity. Is what is happening with the advances of the internet a reflection of the evolution of our collective-human nature? The notion of people having a feeling of belonging, to a country or a community or any sort of group is not new. The famous social neuroscientist Cacioppo and science writer William Patrick traced the evolution of this need, showing how for our primitive ancestors, survival depended not on greater brawn, but on greater mutual commitments[6]. So, people have had plenty of time to create bonds and form ways of connecting to each other. Throughout many generations until today, people have been learning to communicate their behaviours, creations, emotions and thoughts to each other. The same neuroscientist discovered that prolonged loneliness can be as harmful to health as smoking or obesity. His work also demonstrates that social connection has a therapeutic power of healing or preventing some diseases. What does this mean for Meta Products? Well, as Theiner expresses: "The web is simply one of the most recent and socially significant manifestations of people's perpetual drive to become more connected."[7] Meta Products will exponentially multiply these manifestations through new ways of connecting. You as a designer can provide the right impulse to help people manifest their aspirations to connect in meaningful ways.

"How the increased connectivity gives a true advantage to the user instead of adding stress and complexity is where the value of all this lies." — FEDERICO CASALEGNO

People are becoming part of many networks, usually driven by their personal interests or aspirations, but also because 'something magic' happens when a network is in action. We like to simply call it 'something magic' because we cannot really prove what happens or why. Back in the 60's, Peter Blau recognized that social structures have emergent properties not found in individual elements. Later, other scientists, like the same Theiner or Clark & Chalmers[8], also recognized similar things in their studies, claiming that groups have the potential to 'think' in a way that no individual member can or might even be capable of. Opinions vary of course, and scientists have

Innocentive is another example; they call it InnoCentive's Global Solver Network and it is a service whereby companies can easily tap into the talents of the global scientific community for innovative solutions to tough R&D problems. Another great example is Current Media. They broadcast programming created by the audience. Of course there is plenty of support available from advertisers due to the valuable insights into consumer behaviour they gain by letting people programme what they want to watch. "I think the future is more about video on demand, and watching content when you want to watch it," says Paul O'Donovan, of technology research specialists Gartner for the Tech Blog (May 15, 2011)

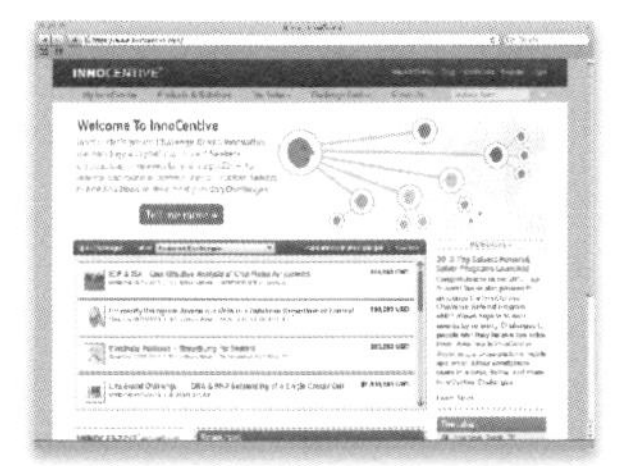

INNOCENTIVE'S GLOBAL SOLVER

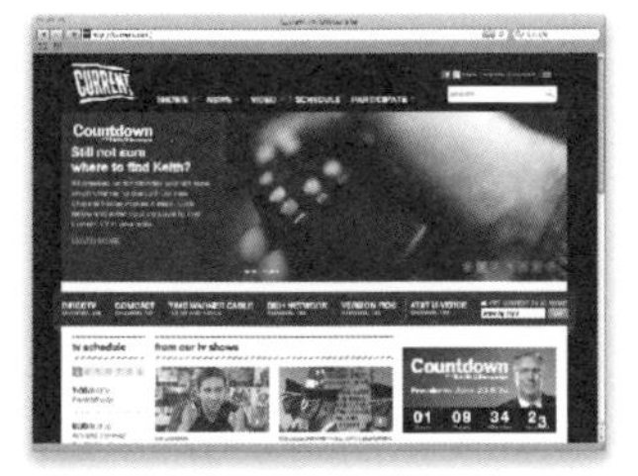

CURRENT MEDIA

to make all sorts of validations on these claims. But we cannot deny that the collective magic is there! A sign of this magic is what's happening at the moment. Just look at the results of Wikipedia — the work based on the motivation of thousands of volunteers to create the world's largest encyclopaedia. The interesting aspect here is that information is now accessible for everybody, thereby making not only sharing accessible, but also revising and updating. A whole new concept of achieving reliability of information was born.

Today, innovation is more a human process than a technology process. Not that it wasn't human before, it's just that now there is an 'old way of doing things' that we may want to change, and where in our quest for progress we were stripped of our basic nature and forced to be standardized, efficient and isolated. Just think of any major technological innovation of the last century that has changed people's lives. The TV, the mobile phone, the personal computer... all of them great inventions that led to new behaviours under the banner of being progressive. However, today, with the advances of technology, mainly digital and information technologies — we are slowly realizing that it can be different and that our lives are much richer and better with diversity and uniqueness, with teamwork and crowdsourcing, with exploration and creativity. At least in some aspects of our modern society. We are also beginning to realize that the old way is not sustainable and it is not human-friendly. A simple example of this is what we have done to our food. We have standardized food production processes and made them efficient. It just seemed like a good idea at the time. We have applied the old way of doing things, in the field of economics and technology, to everything we eat, and the result is that you, a "progressive" citizen of the world, are rarely able to enjoy the luxury of consuming a natural-grown tomato without pesticides, chemicals or hormones. That might be then just the way it is, and people are getting used to it, but that doesn't mean we like it. In the past, we had no power to do anything about it, but now we can, and a sign of this is the application that IQ Advanced of San Diego, CA has developed. It is called iScan My Food, an app for scanning food ingredients and containing a database of information on harmful food additives, toxic ingredients in food and genetically modified foods. All you need to do is take a picture of the ingredients listed at the back of the packaging and the app tells you what you will be eating. Not only that, the app also allows consumers worldwide to submit additional additives or ingredients found in their food products to the database, which will be added to a following update, as new additives are constantly being developed. This is a sign of a new way of doing things.

Coping with multi-user or collective interactions can become complex processes, and form quite a challenge for designers. We could call these 'collective Meta Products' and they can only exist thanks to the collective information, or the emergent knowledge we spoke about in chapter 2, generated by multiple users. The way these information streams flow and how exactly they become emergent knowledge is the subject of many unresolved issues (like privacy and security). But also the emergent knowledge that the interactions you design can create, adds an unpredictability layer on top of your design decisions. What kind of emergent knowledge would your interactions create?

We love this quote by Albert Einstein: "A human being is part of a whole, called by us the Universe, a part limited in time and space. He experiences himself, his thoughts and feelings, as something separated from the rest - a kind of optical delusion of his consciousness. This delusion is a kind of prison for us, restricting us to our personal desires and to affection for a few persons nearest us. Our task must be to free ourselves from this prison by widening our circles of compassion to embrace all living creatures and the whole of nature in its beauty." This quote makes us reflect on whether the paths we've taken as humanity have taken us further apart from each other. Instead of embracing our nature and growing from it, we have slowly alienated it. We truly believe that you as a designer can help people to set new paths that create ways to a more natural and meaningful way of living. This may sound like way too big a task for you, in the end you're just a designer right? Indeed, it is a major task, but we really believe this is precisely the time for you to scale up and bring on the magic!

ALBERT EINSTEIN

3.4 CHANGING OUR DESIGN PROCESSES

Since the Industrial Revolution we've been occupied with changing the structural composition and location of things in order to give them new purposes. In the 19th century, we invented the steam engine and used it to power our boats and trains. Ours is a long history of doing this kind of thing. It seems like 'all of a sudden' we have some new 'material' that potentially can be embedded into everything: bits that co-exist within things, bits that empower things, bits that bring new languages to things and ultimately new meanings to people. Furthermore, whatever you do with bits is potentially scalable and reproducible with zero transaction costs, maybe we could call it an 'easier' way, if we compare it with the difficulties and risks involved in scaling and reproducing real materials. Bits are the basic unit of information in computing systems that have two-valued attributes (yes/no, 1/0, true/false). The physical storage (laptops, phones, etc.) of these attributes are independent, as they can be used for different and multiple purposes in the same device. So a smartphone can be a planner, an alarm clock, a music player, a traffic advisor, and so on. This is quite normal these days. But if you really think about it, this is quite a different concept from what we were used to: a material, like wood or metal, always has a unique spatial location determined by its given structural composition. The combination of both also defines its purpose.

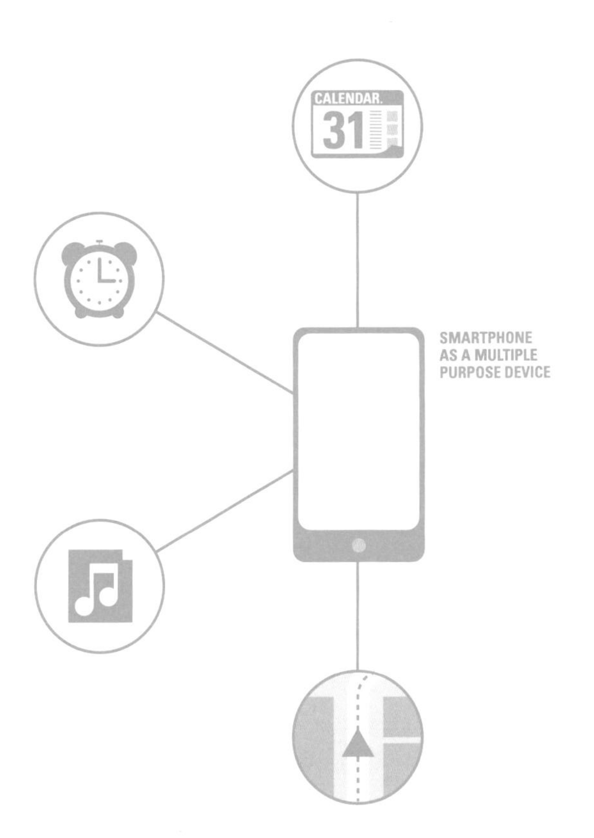

SMARTPHONE AS A MULTIPLE PURPOSE DEVICE

Like a man and a woman in a marriage, bits and atoms are different, but they can live happily together if they accept the differences and understand each other. You as a designer, are the marriage counsellor. In that sense, you must understand how bits and atoms 'think', and know how they behave and what they are good at. So far, you might be thinking that you're already good at this. And you'd probably be right; we're not saying design processes should change, at least not in their abstract sense. Design thinking has always been quite a good basis from which to solve complex problems, regardless of the materials available. However, we believe that some interesting answers can arise if designers ask themselves the question: what kind of impact on my design process does the potential of enabling things with bits have? What is the impact of focusing on the connections between things? What is the impact of 'designing networks'? This of course is your own quest. But we can share with you what we have reflected upon. For us, designing Meta Products has two types of impact on our design process: 1. stakeholders' motivation & quick-testing 2. transdisciplinarity & specialization.

Stakeholders' motivation & quick-testing

To design Meta Products, we basically started a journey to change our design mindset, we call it Network Focused Design. As you might guess, the focus is on the network. We will explain all this more in depth in chapter 5, but now all you need to know is that if you have a network mindset, the connections between the stakeholders that create the network are very important, as well as the information that feeds those connections. What do we mean by stakeholders? Well, they are people, and by 'people' we mean end users, but also companies and organizations that have a relationship of value. So, what we noticed is that sometimes just having a 'great idea' within your team is not enough to make certain Meta Product projects materialize; you need motivated stakeholders to create the network. The motivation of every stakeholder depends on the value they see in the connections they make use of within the network. A challenge here is to anticipate the motivators of all the stakeholders. For this, designers might find it handy to exercise their system mapping skills and to implement quick-testing techniques that include co-creation or active collaboration and communication with the stakeholders.

"Methodologies are good and sufficient. But, what happens if the design teams, which are not always formed by only designers, get pressured by time and budget constraints? These design teams have learned at university that they have to do usability tests, context mapping, ethnography, and so on, but if the company has to achieve a certain number and reduce costs in one week, there is no time or money to implement the methodology learned at university. So these methodologies are not realistic in practice, in most cases. In this sense, practitioners figure out quick and dirty ways to do user tests and all the fuzzy front-end techniques. So, the practical side of all these methods is missing." — JAN BUIJS

Transdisciplinarity & specialization

Design disciplines are becoming more and more strategic. We believe this just had to happen. Design thinking is a relatively new way of doing things if we compare it to strict sciences or other ways of doing things like mechanical engineering, computer science or business disciplines. Design disciplines evolved through recognizing that for some activities it is useful to be able to see the big picture, to deal with fuzzy and immeasurable information, and to understand the problem while solving it. Designers place the user at the centre and make use of design tactics to make educated guesses about what the user is willing to do or not. For Meta Products, designers have to do the same but also taking into account multiple users or stakeholders. Knowing what people are willing to do or not is becoming more and more important in some industries. Hence, designers are gradually being recognized as the new strategists of product-service development teams.

Following on in this line, the strategic skills of designers are all about mastering transdisciplinarity. This term sounds somewhat complex, but it's the only term we've come across so far that, at least on Wikipedia, refers to what we are trying to say: "a research strategy that crosses many disciplinary boundaries to create a holistic approach" and it goes further: "what sets transdisciplinary studies apart from the others is a particular emphasis on engagement, investigation, and participation in addressing present-day issues and problems in a manner that explicitly destabilizes disciplinary boundaries while respecting disciplinary expertise. They are built around three key concepts: transformative praxis, constructive problem-solving and real-world engagement." Design activities intersect with other fields and it is precisely this intersection that is useful for finding innovative possibilities. For instance, some years ago, design and business were speaking completely different languages. Nowadays, design and business thinking are understanding each other and crossing each other's boundaries. For example, Business Model Generation by Osterwalder and Pygneur is one of the first business books that refers to "designing business models", but it is not limited to using the term in a theoretical way, it also applies pure design thinking to create new ways of doing business. Through the main discourse of that book it is shown how the user-centred principle in most design disciplines is applied and mutated into the business world: "No business plan survives the first customer contact." Moreover, the authors recommend "getting out of the building" and testing to validate with the customer. This hands-on iterative process, so common for designers, is now valued in the conducting of business. As a result, designers get involved in business thinking as well. On the Philips website you can find a statement alluding to this: "...designed around our customers. Design has become a marketing tool. It plays a role throughout the product development process, all the way from research to manufacturing to marketing." Whether this happens in practice we wouldn't know, but the fact is that companies such as Philips are integrating design activity and design principles into their core strategies.

BUSINESS MODEL GENERATION, BY ALEXANDER OSTERWALDER AND YVES PYGNEUR

"The cost of any innovation or development is very often the greatest stumbling block. Emergent technologies and their fast adoption are clearly calling for new business models. Leaders must engage a strategic mindset when preparing for this challenging new environment. They must hold a people-centric view of the user not only as a passive spectator and consumer but also as an active inter-actor, creator and producer. Therefore engaging a whole brain vision in the infant concept and process stage is vital. Companies must develop products and services connecting the scientific and social dimension in a relevant way considering the emotional dimensions and its impact and relevance to people's value universe. What I am talking about is to connect people to what really matters." — ANNE LISE KJAER

Design as a transdisciplinary activity will be increasingly recognized as such because organizations need professionals to simplify the many processes involved when making products and services while addressing people's aspirations. We believe that with Meta Products this need will just keep on growing.

"The impacts on design education include: enabling students to learn how to understand and digest research in several fields including design, management, social sciences and IT; working in cross-disciplinary, cross-organization teams; working as leaders and facilitators; learning how to use designers' tendencies to generate novel ideas and novel research methods effectively within teams and projects; and generally being a lot more modest about what design is or can be, and more attuned to the consequences of designers' decisions." — LUCY KIMBELL

Strategy and business together form one of the sides of the transdisciplinary coin of designers, with the other side being the ability of designers to look at people and apply agile social research to acquire valuable design insights. Christine Wasson wrote an article called 'Ethnography in the Field of Design'[9] where she addresses an audience of anthropologists to observe that industrial designers have always tried to meet the "needs and wants" of people using products. However, it was precisely during the rise of Human Computer Interaction (HCI) in the 1980's when anthropologists began to demonstrate how ethnographic investigations into technologically connected communities could help designers better understand the needs of new technology users. They discovered that ethnography was inherently more practical — not by investigating what consumers say, but what they really do. This approach proved to highlight the discrepancies between the designer's intended use of a product and the real behaviour and relationship of people with that product. Later on, the term 'Participatory Design'[10] appeared, involving users in evaluative research: testing existing products or prototypes of developed concepts.

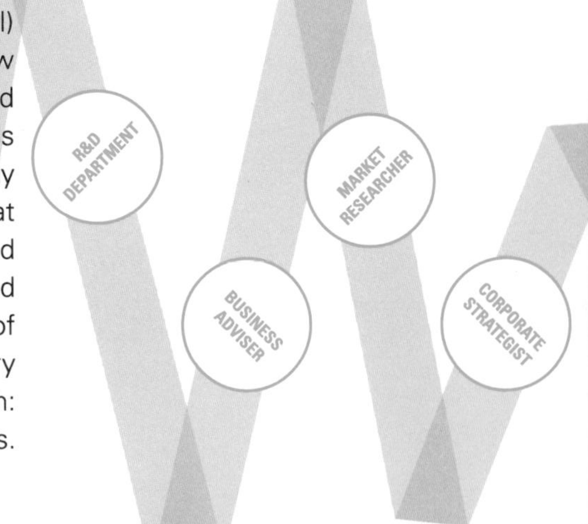

More recently, designers are involved in exploring contexts. Context mapping[11] for example, is a generative technique that intensively involves users in the creation of an understanding of the contexts of product usage. This technique can inspire and inform a design team in the early phase of the design process. This early phase of the design process is what is usually called the "fuzzy front end." Naturally, in this front end, many other disciplines take place and generally these are responsible for setting up a business case: R&D departments, business advisers, market researchers, corporate strategists, etc. Design thinking seems to be the 'glue' between the disciplines, because it allows the interpretation of all the transdisciplinary work into design concepts. Using proper communication tools is gaining in importance as interactions become multidimensional (various time dimensions, multi-users, parallel tasks, automated response, and so on).

To design web-enabled product-service networks, designers need to be strategists and transdisciplinary-skilled. This doesn't mean that you have to forget about the special skills you've been carefully polishing. On the contrary, expertise and specialization is what makes the strategy and transdisciplinarity actually work. You can read more about transdisciplinarity and specialization in chapter 4, where we describe the different types of design specializations required to design a Meta Product.

DESIGN THINKING.

THE
FUZZY
FRONT END

3.5 BUILDING PLATFORMS

Let's take a look back into the greatest technological innovations of all times. Machine-based manufacturing, followed by steam power, the internal combustion engine, and finally electrical power generation. All of them led up to the Industrial Revolution in the late 18th and early 19th centuries. Every single human in the world was touched by this revolution and through this, the history of humanity changed. As Harold Perkin, a British social historian (1926-2004) observed: "the Industrial Revolution was no mere sequence of changes in industrial techniques and production, but a social revolution with social causes as well as profound social effects."[12] Why did the Industrial Revolution come about the way it did? Why was it truly innovative? Well, there was the influence of thinkers at the time such as Alan Smith, Diderot, Voltaire and Marx, among many others, who embraced and promoted the aspiration of mankind's progress, and society was eager to see the tangible manifestation of such knowledge waves not seen before since the times of the Reformation or the Scientific Revolution (14th century). Today people are eager to see their aspirations manifested not in the products they can buy but in what they can do and feel.

The Industrial Revolution offered a platform for broad capabilities for continuous development. Back then, the only way to achieve the progress so greatly desired was to understand that ideas that work in isolation are not always strong enough, and that even ordinary ideas integrated into a whole can be collectively powerful. An empowerment platform where companies, industries and organizations could create more and more products and seek sustained growth by selling them to us. It was only a matter of time before this growth reached new levels that would again change the history of humanity: the internet. A platform where there was space for even contradicting ideas, and where the capabilities for continuous development seemed greater than ever before; democratizing knowledge and transgressing geographic frontiers to reach other people and their information… a whole new world where not only companies and governments but also you and me, and everybody else could actively be a part of. We don't have to tell you the rest because you are experiencing it every day, but we can say that the internet is potentially the ultimate technological innovation capable of creating more platforms to express, to learn and to understand and shape ourselves and our world. So, what we can see is that the Industrial Revolution and now

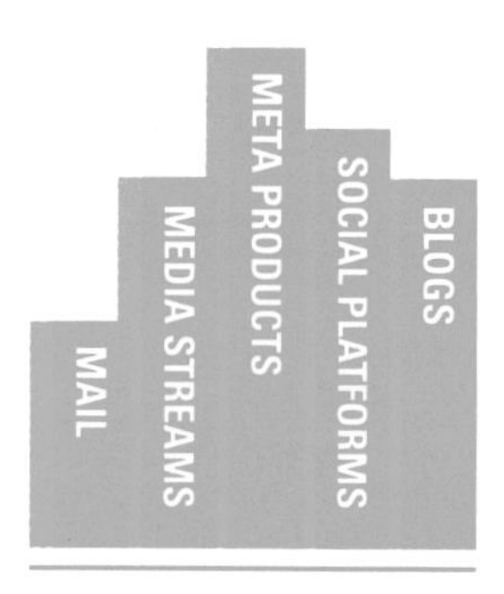

THE INTERNET REVOLUTION PLATFORM

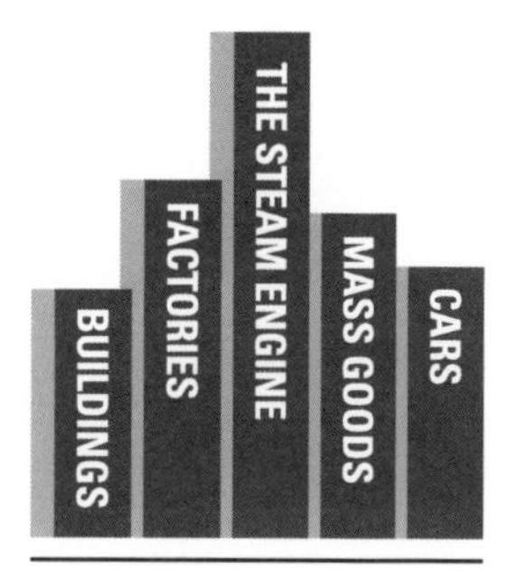

THE INDUSTRIAL REVOLUTION PLATFORM

BRAND-DRIVEN INNOVATION, BY ERIK ROSCAM ABBING

the internet are platforms with apparently unlimited ambitions. Both set suitable conditions and systems that enhance our capabilities and give us resources for particular environments. Both are socially pulled and include many elements conforming to a meaningful whole. However, the difference is that in the Industrial Revolution the atoms were the subject of most value and therefore the focus of our attention. Now, the atoms (the tangible material or products) are just an element in a network of value where the connections — virtual or real — are the focus of attention. Perhaps this is why we think it is so popular now to talk about service design and service-innovation models. In a conversation with José Laan from Syntens, she explained how companies seek to make the transition from 'product-oriented' to 'service-oriented'. Syntens teamed up with two Dutch universities and a user-centred design bureau to create a service-innovation model to help particularly small and medium enterprises to make this transition[13]. The model proposes seven steps. The first ones are basically a fuzzy front-end of innovation development that leads to a promise and a strategy to accomplish the promise. It looks a lot like a design process adapted to the business-minded. It is iterative and customer oriented, and it aims to help companies build their own service platforms with flexible elements that can adapt to the needs of the customer. Erik Roscam Abbing, with his book Brand Driven Innovation[14], shows how today's companies are looking to create networks of value and building platforms of growth. His book helps you to isolate the old-fashioned roles of products and services inside companies and to focus on 'the brand' as the glue that binds it all together, inside and outside. We could call it a 'brand platform' where the customer, the company and all stakeholders interact to create their network of value.

"Meta Products mean a larger scaled 'living environment'. The designer must manage this complexity in order to develop lean interfaces, platforms and emotional hardware that appeal to people's diverse needs and personal relationships in an information context." — ANNE LISE KJAER

So far, the platform of internet has allowed an ever-growing digital representation of our world, our dreams and fears, our right and wrong… and it has even bigger ambitions. For instance, the Echo Nest platform is the world's first machine-learning platform for music.

From the Echo Nest website: "…the Echo Nest's music intelligence platform combines large-scale data mining, natural language processing, acoustic analysis and machine learning to automatically analyze every song on the web to extract key, tempo, rhythm, timbre and other attributes — to understand every song in the same way a musician would describe it (i.e., "swing groove, tempo = 100 BPM, 4/4 time, key of B Flat, mezzo piano"). Echo Nest analyzes every blog post, every review, every piece of text that describes music to understand in real time what the world knows about every artist, release and song."

This is a good example of today's ambitions to retrieve and manipulate information and knowledge in better, ubiquitous, real-time and mobile ways, and this can only be done on top of proper platforms that can help to scale up the individual ideas into collective action. What is interesting about all this is that today we can actually see individual ideas, small and medium enterprises creating platforms. In the past, you had to be a large corporation to be able to make something innovative. Nowadays, thanks to the web and ubiquitous technologies, you can reach more people and try out services in shorter time-frames. But of course, there is always another side to the coin: finding the value of your network and the best way to scale it up is not that easy. The challenge here is to identify the best relationships and to have motivated stakeholders. In chapters 4 and 5 we will talk more in depth about this.

"We do research into scalable models for ICT services. Together with partners in industry, governmental or non-governmental organizations. We apply these models as they already have a customer base, infrastructure or a network. You can scale up services, but you can also focus on scaling up the impact with these services. It means that the scale of the overall impact can be larger than the scale of the service itself, because it can change or influence the possibilities, mindsets and behaviour of other organizations or companies." — JENNY DE BOER

THE
ECHO NEST
PLATFORM

CHAPTER 3 — SUMMARY INSIGHTS

1 The more mobile the internet becomes, the more it acquires the ability to become embodied virtually anywhere and in anything. This is resulting in products, ideas, knowledge, services, people, and environments communicating in new ways enabled by the web and other ubiquitous technologies.

2 Products and services will never be stand-alone anymore. They will be increasingly related to other products and services and to other things, people, and environments via the information they generate.

3 Meta Products provide a special focus on the connections that build up a network rather than on the isolated product or the particular service.

4 If you design Meta Products, you are, roughly speaking, designing networks with connections that switch from digital to physical and vice versa. All this switching may mimic our 'real' relationships, improve them in some way, or create new relationships, and — ultimately — new behaviours.

5 The complexity of a human experience may become even more intricate when the possibilities of connecting and communicating multiply. This results in new languages and new interactions yet to be refined until the meaning is clear for the people engaged with them.

6 Think of a Meta Product as a conversation between physical things, information, people and spaces (and whatever else you can imagine) enabled by the web and other supporting technologies.

7 Use your most sophisticated empathy skills as a designer to try to understand the relational process that takes place when people assign meaning to new interactions.

8 Trying to understand mental models may help you to design interactions that are meaningful for people.

9 All human processes change or evolve during time, even when the interactions remain the same. Because people learn and create different associations with the same things they have been interacting with, as a designer you have to be aware that the relationships in a network are perishable.

10 The motivation of stakeholders depends on the value they see in the connections they make use of within a network. A challenge here is to anticipate the motivators of all the stakeholders. For this, designers might find it handy to exercise their system mapping skills and to implement quick-testing techniques that include co-creation or active collaboration and communication with the stakeholders.

11 Design disciplines evolved through recognizing that for some activities it is useful to be able to see the big picture, to deal with fuzzy and immeasurable information, and to understand the problem while solving it.

12 This hands-on iterative process, so common for designers, is now valued in the conducting of business. As a result, designers are becoming strategists.

13 In the Industrial Revolution the atoms were the subject of most value and therefore the focus of our attention. Now, the atoms (the tangible material or products) are just an element in a network of value where the connections — virtual or real — are the focus of attention.

14 Today's companies are looking to create networks of value and building platforms of growth.

15 Today's ambitions to retrieve and manipulate information and knowledge in better, ubiquitous, real-time and mobile ways, can only be done on top of proper platforms that can help to scale up the individual aspirations into collective action.

16 An important challenge when designing web-enabled product-service networks is to identify the key relationships and to have motivated stakeholders.

SONGDO INTERNATIONAL BUSINESS DISTRICT

Songdo International Business District (IBD) is a city being developed on roughly 6 km2 of reclaimed land in South Korea along Incheon's waterfront, 55 km from Seoul. According to developers Gale International and POSCO E&C from Korea, the metropolis will be a new benchmark for sustainable, city-scale development and innovation. International architects Kohn Pedersen Fox are responsible for the design of commercial office space, residences, retail shops, and hotels, as well as civic and cultural facilities. In 2015, when Songdo IBD will be the new home of 300,000 inhabitants, everyone will be plugged into the city's own giant network. In other words, the entire city can be regarded as a huge Meta Product.

Cisco, the manufacturer of the electronic data grid, has said the city will literally run on information. Everything will be connected to the grid, thereby making it possible to automate the lives of the inhabitants in entirely new ways. Health monitors will be placed in the workplace, computers will be integrated in every house and in the streets. Ubiquitous computing, as part of the smart infrastructure of the city, will be everywhere.

The large-scale implementation of ubiquitous computing is not only technically driven, care for the environment is equally important. Once it's completed, the whole of Songdo will produce a third fewer emissions than current cities of comparable size. For the field of Meta Products as a whole, Songdo is more than interesting, because it's a huge testing environment of possibilities. Potentially the city will provide us with information, and adapt instantly to our needs. It will be exciting to watch the results of people living in such a networked city. Failures will probably teach us a lot, since so many new interactions will be possible. Maybe the aspirations of future generations will lead this project into a direction not yet plotted by the architects. If you think Songdo is a one-time experiment, think again. Gale International is planning to build another 20 connected cities in China within the next 10 years. How do you think living in a connected city will differ from living in a traditional one? Which one will fulfil our aspirations? The fact is that the potential is there: cities will be living entities and will merge with us in everything we do. For you as a designer, this opens a lot of opportunities that will require your fullest empathy and reflective skills to design meaningful interactions.

IMAGES PROVIDED COURTESY OF GALE INTERNATIONAL, LLC.

EPYON POWER

Epyon Power is a Dutch company, pioneers in electrical vehicles (EV) with fast battery charging. They offer two key products: a charge station and an accompanying web application. Typical charging times are between 15 and 30 minutes, making the product suitable for business fleet owners, as well as utility infrastructure suppliers. Epyon operates mainly in the utilities, transportation and infrastructure branches.

While the charge station is the core of their business proposition, Epyon has built a network of value around it enabled by a web-based control, management and maintenance system. For example, the web application allows a charging infrastructure operator to access status information and statistics from the chargers at their sites, including kWh consumption and statistics on a weekly, monthly or yearly basis. Additionally, an operator can use the application to configure chargers. This remote access and control of their electrical chargers is only possible because the chargers are all connected to a network. Epyon provides a suite of APIs which enables the chargers to interface to third-party billing servers, fleet management systems, smart grids or demand-response applications. In other words, they provide a platform of products that are accessible via their network.

As a developer, you can extend the functionality of the chargers by letting your own programmes talk to their APIs. For example, imagine a transportation fleet that consists of multiple electrical vehicles. Each of them continuously transport goods in a certain controlled environment. Electrical chargers are placed at strategic points, allowing the vehicles to charge when needed. But only one charger can be used at a time. A developer writes an application that monitors if a charger is currently being used or not. The application could then send vehicles with almost drained batteries to the most strategically placed charger, taking into account all other vehicle's batteries too. This way, the application maintains a high-level overview of the complete transportation fleet. It is able to optimize the whole network because of the available touchpoints of relevant information.

Epyon is an example of a company that provides not only a product, but also a platform. This way, customers or developers can easily customize and embed Epyon's services in their own organization according to their particular needs.

epyon POWER FAST CHARGE
RESTAURANT SHOP
FAST

RESTAURANT
FAST

epyon POWER FAST CHARGE
SHOPPING MALL
FAST

How would you define a meaningful experience?

JEROEN VAN GEEL
Fabrique

A meaningful experience has all to do with the context. The more a product is able to understand the context in which the user will interact with it, the more meaningful an experience will be. Awareness is also very important, to acknowledge a meaningful experience as such, the consumer should be aware that he is experiencing something. Design can provide the tools to do so, for example, storytelling is a good way to make experiences more meaningful. Imagine that a product or service talks to you "like your father" instead of like a stranger that wants something from you, you would feel identified and the experience would be more meaningful to you. This sort of tool, and many others we can provide to help consumers have meaningful experiences.

What kind of challenges do you see for designers when designing Meta Products?

LUCY KIMBELL
University of Oxford

It is a delusion that any designer can convey the "right" meaning or design an experience to play out a particular way. What we know from studies of media and other kinds of consumption is that people bring their own desires, identities and purposes to their activities which no designer can foretell, however much research they think they have done. What designers can do is assemble things into temporal and spatial arrangements with an understanding of how things might perform — choreography and curating are useful concepts here — but with the modesty to understand that people bring their own meaning-making to their encounters with such assemblages in practice. The implications for designers are (a) making sure they work with sociologists and anthropologists who have useful concepts, tools and approaches for trying to understand and describe what people are doing in their engagements with such products, services and spaces, and what this might mean at collective rather than individual levels; and (b) having iterative design processes that allow this developing and emerging understanding of what things might mean continue to shape design, testing, development and delivery.

ANNE LISE KJAER
Kjaer Global

As globalisation and mass consumption make the world increasingly homogeneous, people's search for meaning and emotional connection intensifies. Computerised mobility and social networks are the opportunity to create our own unique identity and discover a new world. We want the real thing, be it food, travel or art, and happily mix it up into our own interpretations and personal narrative. 'Real life' experiences providing involvement, inspiration and informed knowledge in a fulfilling manner will be the 21st century brand fabric.

ALEXANDRA DESCHAMPS-SONSINO
Designswarm

An experience is something out of the control of the designer; you cannot actually design an experience. An experience is felt, lived and we can only guess it. An interaction is something you can aim for and that you can actually design. The underlying principles around both are very very different, the role of the designer is very different in each of them. Some claim that they can design something to experience this or that in a space for example, but actually that experience is out of your hands, because maybe the person interacting with it is very sensitive to sunlight or whatever other thousand things. Designing an interaction is the most empathetic way of guessing what people might experience.

JAN BUIJS
Delft University of Technology

I see the challenges for designers more on the manufacturing or processes level than in the design mindset per se. It is all about scale and size we are dealing with, it is not necessarily different what we are doing now but it just comes in a growing scale and a larger size when we design systems or networks.

MADDY JANSE
Philips Research

I think the first idea of the telephone was to bring classical music to the masses. That was the original idea of running out that new technology, and you know what happened, it became something totally different. So it takes a couple of steps before the real applications are being taken, and sometimes they can be totally different than the technology people or the designers had thought at first. In 1988 it was predicted that in 2000 we wouldn't have a TV because we would do everything through the internet. Sounded good but it didn't happen so. You have to make a distinction between what technology makes possible and how social developments happen to adopt the technology.

CHAPTER 4 — *The Perfect Meta Product*

perfect Meta Product / valuable relationships / information-fuelled / learning / new knowledge / new behaviours / exchanging resources / scalable platform / motivated stakeholders / atoms' business models / bits' business models / collaboration / active links / design teams / system / product / interaction / service / information exchange / smart devices / new communication skills / mapping motivation / exchanging resources / social capital / leading motivation / tracing the impact / accountability / decreasing complexity

4.1 REFLECTING ON IDEALS

Aristotle defined perfect as something so good that nothing of its kind could be better. So how can a disruptive innovation such as Meta Products be perfect? Well, perhaps perfect doesn't exist and if it does, the concept would be perishable. In any case, there is always a moment in which we have a perception of how something could be so good that nothing else could surpass its greatness. Then of course this perception changes. We've ideals about our jobs, our hobbies, the stuff we buy and basically about everything we do. As a designer, you have ideals as well, and it is important to reflect on them to help you not only to optimize your design process but also to be conscious of the impact your design has on the world. In this chapter we share our ideals about designing Meta Products. At the moment we could say that a Meta Product can be perfect when it builds valuable relationships or connections between the members of a network of people, their resources and their context. This will probably hold true for service design and other types of design activities that might not involve any technology at all. But our reality today is that as the web and many other ubiquitous technologies improve and become affordable, they can be potentially embedded basically anywhere. This means that many of the valuable relationships you can think of between people, their resources and their context will, one way or the other, be fuelled by the information enabled by the web and ubiquitous technologies. This has — and will continue to have — a huge impact on innovation in various industries where new ways of generating, reaching, sharing and manipulating information will lead to new valuable relationships. And along with them, undoubtedly, new products, new services and new opportunities will come along. But how can we identify these valuable relationships? How can we achieve true meaningful design for our connected world? We've defined our ideals about it, a sort of guidelines with some principles to keep in mind which might help us design close to perfect Meta Products. We present these ideals here because we believe it's important for you to prepare your mind. Most importantly we would like to encourage you to think of your own ideals before actually starting to design Meta Products.

4.2 ENCOURAGING MEANINGFUL EXPERIENCES

You can get quite a lot of inspiration from learning and cognitive theories when trying to understand how someone can live a meaningful experience. Distributed Cognition, for instance, is a learning theory that states that when you learn, you model your own world-view. And you do so in a process that involves your own prior knowledge, the present experience, and the situation itself. This is how we learn, or in other words, this is the way we assign meanings and create new knowledge. Of course we learn in different ways depending on whether you're a famous singer or a dentist, a rebel teenager or a just-married couple, a steel company or a financial consultancy, but the basics are the same. Now, imagine that your Meta Product enables new streams of information. How would this new information impact people's learning behaviour? Moreover, theorists say that we learn better when we are motivated. Motivation is the driving force through which we achieve our aspirations and it is closely related to the way we assign meaning to everything we do.

"A meaningful experience has all to do with the context. The more a product is able to understand the context in which the user will interact, the more meaningful an experience will be. Awareness is also very important, to acknowledge a meaningful experience as such, the consumer should be aware that he is experiencing something." — JEROEN VAN GEEL

To use the term 'meaningful experience' is always a bit vague, but that's probably because we are referring to an effect that is very personal in individuals and in groups. We as designers can only try our very best to guess right and to encourage it. When you get to design a Meta Product, keep this ideal in mind and improve your empathy skills, including understanding the collective behaviour and basic cognitive processes of human perception. In chapter 5 you will find more hands-on recommendations on how to design interactions that aim to render meaningful experiences.

"...the role of designers in Meta Products is to match the technology and the aspirations of people."
— JAN BUIJS

So, from learning theories we might infer that when you live a meaningful experience — for example during a music concert — you are actively learning and fitting into your world something that helps you both recognize and achieve an aspiration. It could be a direct aspiration at the very moment you are living the experience, or an indirect aspiration that relates to something else in the long term. As Nathan Shedroff says, 'The most successful experiences are meaningful'[1]. And probably this is what we want to achieve when we start out to design something. For example, imagine a Meta Product dedicated to the safety and security of a hospital. Let's say it consists of a network of devices such as a fire alarm that senses smoke and immediately sends signals to the closest available Fire Department. There are also image and movement sensors all around that detect people and autonomously determine whether to allow or block access to certain places depending on whether the person is recognized as a nurse, a patient, a doctor or a visitor. Every medicine would have an ID tag attached to keep control of the inventory and to track the journey of that medication from A to B. There would be interfaces providing this information when necessary. Some interactions would be automated and others would be controlled by the hospital staff. Now think how this new information would change people's behaviour? Are there new interactions that help them identify and reach their aspirations? Whose aspirations? Would people fit this type of Meta Product into their lives?

4.3 BUILDING NETWORKS OF VALUE

Today, your services, products, people and spaces can 'team up' by using all sorts of information streams to offer you personal and relevant uses. This is possible thanks to the web and ubiquitous technologies. But just because something is possible doesn't mean it's valuable. 'Encouraging meaningful experiences' is at the centre when designing a Meta Product (the people or the users are at the centre of the design process). The way these experiences are orchestrated, or in other words, the way to make them happen, is by building networks of value. This means creating ways to motivate the members of a network to exchange their resources. During this process, new members, new resources and new methods of exchange might arise. The information stream would be 'the fuel' of it all.

By 'members' we mean the consumer, the company, and everyone that could actively exchange resources. By 'resources' we mean services, products, concepts, money or know-how.

So how can we build networks of value? How can we build a motivated network? Let's take an example. Pachube is a platform that stores, shares & identifies real-time sensor, energy and environment data from objects, devices & buildings around the world[2]. That sounds fantastic, but of course to make sense of all that and come up with a sustainable business is quite complex. For instance, after the earthquake and tsunami in Japan on March 2011, a developer created an Android app that shows the radiation from the Fukushima nuclear power plant and visualizes the wind of Japan. It also shows the distance between the power plant and your location on the map. Pachube makes it possible to show numerous measurements by Geiger counters owned by citizens. How is Pachube building a motivated network here, or a network of value? We guess that first of all, the service is meaningful to the people in that context; secondly, Pachube is trying to build a more robust, compatible and scalable platform where people, organizations, service providers, objects, products and many other things can become part of a network. The more robust, compatible and scalable the platform is, the more Pachube can bring down the costs of handling masses of real-time data. This in turn will attract more developers of hardware and software and more people to pay the subscription and connect with Pachube.

"In an always-on era we haven't even started to grasp the potential of the internet nor its side effects. The current value on a societal level is set on collaborative tools, increased transparency, open source, real-time sharing, knowledge exchange, social networks and living breathing digital communities. With that we see the rise of dialogue-driven innovation — already impacting business models now but even more so in the future."

— ANNE LISE KJAER

PACHUBE

Building a network of value means providing a service that is meaningful to the people in a network and then offering them a platform to adapt and grow that service as they need it. Right now, you're perhaps thinking about what 'the' business model for Meta Products might be. First of all there is no one answer to that question, because it depends on the motivation of the people in a particular network. But let's try to reflect on different business models that have been proposed in recent years:

People pay for the actual atoms

For most products, consumers actually pay for the atoms themselves. They pay for the materials, the production and assembly processes, the distribution, and the extra margins on production prices. If there is a digital or online part, this is assumed to be free.

People pay for use

In some cases, businesses are selling products on the basis of usage. Particularly if the investment is high, like buying a car, consumers can also rent a car when they need one. People find it convenient to pay only for the time they actually use these product-service combinations.

People pay for recurrent products

Some markets, like the magazine and newspaper market, have a short life cycle, so it becomes handier to just subscribe to receive a certain amount of items within a given time span.

Giving away products for free

Businesses are giving away their products for free more than we might think. In most of these cases, advertising covers the costs of producing and distributing these products. Think of the many free magazines where advertisers indirectly pay for all the costs. This is called the 3-party market (producer, advertiser and consumer).

In the digital domain, other business models can be identified:

Selling licenses

Most software is sold on a license basis. Consumers pay for a certain version with certain features which they can download at purchase.

Selling subscriptions

Like physical products, web services are often sold via subscriptions. Think of your hosting service. You probably pay an annual fee to get your website hosted.

Giving away web services for free

There is an extremely large group offering their web service completely for free, and this group is increasing every day. There are several models that can serve to keep these businesses alive. One is through advertising: let advertisers pay for access to an audience who, in return, pay nothing for the service itself. Another way is merchandising: offer merchandising around the relevant free web service. Services are also given away entirely for free as an investment for growth. Small companies launch and grow a web service until it's big enough to sell it to a web mammoth such as Google or Yahoo. In the meantime, consumers have been using the service for free. YouTube did it. Lastly, another way of giving away web services for free is the popular freemium model: offer a free basic version of the web service, and sell in-app features later on. Many Android apps work this way, though Apple is not fond of this scheme. They seem to think that this freemium model can easily turn into a 'scam', while encouraging the proliferation of low quality apps that are just interested in increasing the number of downloads to attract advertisers. What do you think?

Crowdfunding

This is a very interesting model we see on the increase these days. With this model, consumers can choose to give a donation for the service they use or to support any other type of cause. Web apps and SMS functions are coupled to allow people to pay without any hassle.

IBM SMARTER PLANET EXHIBITION

Meta Products consist of relationships between the web, devices, people and spaces, hence the business models should be flexible and creative. Probably yours will be a play and mix of the above stated business models. Depending on how the network of value is built, you will be able to identify whether selling the physical product and giving away the web service is a good idea. It's a myth to imagine that people don't want to pay for the bits or the web service. In a way, the word 'pay' is already sounding old-fashioned. We believe that people are willing to exchange for value, in whatever form it may come, as long as they are able to see it clearly.

In an interview with Usman Haque, founder of Pachube, he said that Pachube's business model is "slowly adapting as we talk to big and small businesses." Furthermore, in that interview he identified privacy as the most valuable business model: "if you want a free service, then data will be open: but if you want privacy, that's what you will pay for."[3]

In a recent article in McKinsey Quarterly about the Internet of Things, the current challenge of building networks of value is described: "Business models based on today's largely static information architectures face challenges as new ways of creating value arise. Now is the time for executives across all industries to structure their thoughts about the potential impact and opportunities likely to develop from the Internet of Things." Actually we believe designers that are skilful in designing networks will have a crucial role to play in helping industries figure out what to do with these technological phenomena.

In our future connected world, business models will arise from the cooperation and teamwork between industries, for example by building compatible platforms for visualizing, mapping, feeding input and output, manufacturing of devices, software development, and so on. We can see signs of this already as services and devices arise on top of existing data such as the mapping from Google or the social information of Facebook, Twitter or LinkedIn.

"Value is a complex concept that has a multitude of faces, but perhaps in the overall scene of the internet of things there is value in the possibility that it gives us to measure, manage and being told if what we are doing is less efficient than something else. Especially when the effects are on a larger scale, such as the consumption of energy. When this value is recognized it actually changes behaviours." — ALEXANDRA DESCHAMPS-SONSINO

The big internet companies like IBM, Apple, Microsoft and Nokia-Siemens are figuring out how to create web-sensor-networks one way or the other. Wuxi WNS in China has enabled interesting collaborations between companies and the government aiming to build compatible platforms for the web and ubiquitous technologies, in order to make possible what they call a 'smart planet'. They've constructed two 'information service parks' and a 'university park' where information and technology service providers such as Nokia-Siemens, China Telecom, China Unicom — among many others — and universities can assemble and implement their networks for 'smart logistics', 'smart industry', 'smarty environmental protection', 'smart healthcare', 'smart urban transport', and so on.

"...for the potentially valuable developments you need basically standardisation. It's ridiculous that you can't use the same mobile phones in China as in the Netherlands. With computers it's even worse, you barely have an Apple that can talk to a Windows machine and the other way around. Imagine in the Internet of Things, if I have a X coffee machine hanging in my home network and I want to change it to a Y coffee machine, but this one doesn't talk to my home network...it won't work. That's why we need standards and compatibility. This holds for multitudes of different consumer appliances." — MADDY JANSE

So where do you stand as a designer in all this? We believe the involvement of designers is a crucial element at the fuzzy-front-end of value creation. Designers are the active link, translators, mediators and communicators between the processes of creating platforms, building networks and encouraging meaningful experiences. You may not have invested in an industrial park in China or be in the process of developing the ultimate technological solution, but you do understand how networks behave, how people create knowledge, how people become motivated and assign meaning to what they do and design or co-design interactions accordingly, so that technology truly serves us rather than us becoming servant of technology.

"Companies who think innovation is about invention and technology push are likely to struggle with the Internet of Things. In contrast, if they see innovation as situated within socio-material networks they have something meaningful to offer, or what you call Meta Products." — LUCY KIMBELL

There are some intensive efforts being made towards achieving standards and compatibility in the future internet. For instance, you can get a fair explanation of the latest developments on the Internet Protocol Version 6 (IPv6) on Wikipedia: "IPv6 is the next generation of the Internet Protocol that is currently in various stages of deployment on the Internet. It was designed as a replacement for the current version, IPv4, that has been in use since 1982 and is in the final stages of exhausting its unallocated address space. Despite a decade of development and implementation history, IPv6 is only in its infancy in terms of general worldwide deployment. A few international organizations are involved with IPv6 test and evaluation ranging from the United States Department of Defense to the University of New Hampshire. On July 13, 2010, native IPv6 over Universal Mobile Telecommunications System (UMTS)/ General Packet Radio Service (GPRS) was successfully tested in Belgium and The Netherlands within a vehicle platform as an Intelligent transportation system (ITS) solution."

4.4 MASTERING TRANSDISCIPLINARITY

What kind of team would you need to design a Meta Product? Of course you would need specialists in particular domains. For example, if the Meta Product is dedicated to the agricultural industry or to the treatment of Alzheimer, you will surely collaborate with agronomists in the first case and with neuroscientists in the latter case. You would probably need an engineering team with a database engineer, software integration engineer and an application engineer, to name just a few. This book is about designers so we will focus on the design team. In any Meta Product case, you will need a strong team of designers, and these may include service designers, system designers, strategic designers, product designers and interaction designers. All of them, plus all the other specialists, will have to master the art of transdisciplinarity. And this is particularly important if the network is complex in terms of feeding or analysing dense streams of data, multi-users, multi-interfaces and parallel activities, through spaces and devices — in real time. Some definitions exist in literature around transdisciplinarity. The core idea involves "different academic disciplines working jointly with practitioners to solve a real-world problem."[4] A real-world problem in this context means a situation where there are many factors involved of a technical, financial, political or organizational nature. At the same time, various beliefs and perceptions are vying with each other to find some common ground. 'Trans' means across, on the other side or beyond. Transdisciplinarity means that disciplines working together cross their boundaries with the aim of finding new solutions. It's all about interventions between disciplines that are necessary to solve problems that have not been — or cannot be — solved through conventional multidisciplinary practices, or when the aim is to innovate.

"... there's a flipside to the Design coin. This is the collaborative dimension, in other words the social-interactive dimension around the product or the service you are designing. There are many people and many disciplines involved, including different kinds of actors at varying levels and hierarchies. All of them are design actors and you as a designer are intervening in their design processes. This intervention can also be carefully designed. By doing this, designers become some sort of important change agents within organizations. I think that this role of designers will even become more important with the Internet of Things or Meta Products, because products, services and information are becoming integrated into more complex systems." — FRIDO SMULDERS

"...you will see that more and more people are designing but they just don't realize it or they were never trained to be designers. I call it 'designing beyond design'. If you were raised and trained as a designer, you should be able to work with non-designers, I call this Design Acting. Design acting is collaborating with non-designers in a design process. I think this will become very important in the near future." — FRIDO SMULDERS

To give you a better idea of what exactly a Meta Product design team does, here are a few examples:

There are entire journals and books dedicated just to the concept of transdisciplinarity and to finding better ways of collaboration in complex processes.[5] We hope that this ideal encourages you to find your own ways of truly collaborating, getting involved in your team members' processes with empathy, and polishing your communication skills.

The system designer

A system designer would be someone you need right from the start to shape the concept and define its inputs and outputs at an abstract level. Wikipedia identifies System Design as the process of defining the architecture, components, modules, interfaces, and data for a system to satisfy specified requirements. So the system designer would look at the stream of information exchange from a logical and technical point of view. This involves analysing the actual data input and output of the system and working out how this data is processed along the way. A clear idea of how technically feasible the project is will be definitely an important contribution of this designer. System designers that are blessed with a whole-brain thinking cognitive style can consider themselves fortunate, since they will be able to look at the design challenge with both a socio-emotional mindset and a technological mindset. And they will also be able to facilitate and bring about meaningful but feasible connections.

The product designer

A Meta Product will contain devices that will be able to constantly measure, sense, mediate, filter and 'talk' to us in ways they didn't do before. This will call for product designers, who deal with issues such as material characteristics, aesthetics, ergonomics, and mechanics. The designs of most products will tell us something about their functionality, how they work, or how we should operate them. Handles, buttons, the general shape, they all reveal something about how we should treat the product. Sometimes these clues can be small. Let's look at the Polaroid SX-70 for example: it looks odd at first, but once you identify the lens and the viewfinder it's unmistakably a camera. The harmonica-shaped rear end gives you a hint that the camera can be collapsed to protect the fragile parts so you can easily take it with you.

Product designers often apply the so-called MAYA concept (Most Advanced, Yet Acceptable), which means that when designing the most innovative products, you should stick to a few aesthetic features people will still be able to recognize from older products so they can more or less understand what the product is about. The Polaroid SX-70 is most definitely an excellent example of MAYA. The 20th century design trend 'form follows function' has given us products that ultimately show their true nature, and are transparent in all of their functionalities. However, in today's world, where technology no longer constrains the look of a product, and where it's increasingly about the digital contents of a product, we are starting to see plenty of black boxes. Have a look, for example, at the SenseAware, Nike+ sensor, Fitbit or Wattcher; these are all examples of products that have slick designs, but none of these designs reflect what the products are or how they are supposed to work. As a user, you have to learn by exploring the product or, more likely, you'll have to connect it to the web and learn how it works from your computer screen.

POLAROID
SX-70
SONAR ONE STEP

Must the design of a device underline its functionality? In software design, designers often use archetypal metaphors that people recognize and assign a certain functionality to. Think of the postage stamp icon for your mail inbox. Doing the same in the design of devices might not be the answer at all, but the truth is that tangible devices are speaking a different language now and product designers have to realize this.

New product designers are learning as well that the tangible device is part of a whole network of value, and that meaningfulness is a process derived from a combination of touchpoints (digital functions, services, timing, relevancy and so on). The product designers of today are learning to treat data and information as a material, just as Mike Kuniavsky explains in his book Smart Things: information is an agile material that needs a medium[6]. We believe that today's product designers are learning a lot, breaking through paradigms and preparing themselves to cope with the challenges of Meta Product design.

The interaction designer

Interaction design is a multi-disciplinary field in itself, incorporating aspects of psychology, anthropology, sociology, computer science, graphic design, industrial design and cognitive science. The focus of interaction designers is on human experiences. While actual human experiences cannot be designed, we can at least aim to design interactions that encourage certain experiences.

Meta Products will create new interactions, and consequently new languages and ways of communicating. In this sense, interaction designers face a two-fold challenge. In the first place, there will be more "newness" and it will be the interaction designer that can stimulate experiences that help people fit new interactions into their lives. Interaction designers are all about communication, they know that people are willing to learn new interactions as long they can see the value of it in their lives. So, interaction designers are strongly reflective practitioners who study how people perceive, learn and build relationships between their products and services, both individually and collectively within a specific context. The other part of the challenge concerns the increased complexity in communication data. Just imagine millions of sensors generating data about so many things that didn't previously exist. For database engineers, making sense out of all that is a challenge, not only in the technical sense but also in presenting information to people in a way they can easily understand. So, interaction designers will have to considerably improve and expand their visual communication skills to translate complex information into meaningful interactions.

"The classic 'usability' definition is basically operational. For things like how long it takes to do something or how many times something is repeated, it can be very strict. 'Ease of use' goes one step further. It aims to allow people to intuitively use something the first time they see or encounter it. Today, we are more interested in the design of total eco-systems, which contain totally different services, different devices. People might not even use all of them, and maybe just one or two on different occasions. And this is why usability has become rather mundane as a term, and the need arose for a new term to be created, leading to 'user experience'. If you look at the professional domain of user-system interaction and you want to make an 'experience' operational, in other words you want to be able to evaluate it or test it, then you revert to the old usability type of activities. When you're simply designing for usability, the constraints are strict, and sometimes they have to be strict. When you design for 'experience' you have the freedom to be more experimental and creative. Designing for an 'ideal experience' is quite complex, especially when it differs per individual. I think you have to give people options to create their own experience within the environment you provide." — MADDY JANSE

The service designer

The role of a service designer in the design of a Meta Product is basically to make all design activities congruent. The service designer looks at the 'big picture' and will propose design guidelines in a strategic and systematic way. The service designer will look for commercial potential while using strong empathy skills to recognize the motivation in a network. He or she will be able to conduct agile and hands-on ethnographic research as well as trends & market analyzes in order to translate findings into applicable design insights. The service designer will work more closely with the interaction designer, particularly when testing interactions. At every iteration in the process, the service designer will reframe the design guidelines and maintain coherence within the rest of the design team. Organizations and industries are slowly recognizing the importance of viewing the overall service environment and its interactions, rather than just focusing on the tangible result (the product), which has been normal practice for many years. By looking at the overall service ecology and its interactions, industries will be able to understand the contexts within which the real drivers of successful products lie. One of the most inspirational literature sources on service design is the work of Vargo and Lusch[7]. They developed a theoretical compendium rather than an operational one of what is currently happening with products and services.

They refer to it as Service Dominant Logic (SDL) and, as the name implies, it is a logic — or a mindset — that drives the way we do business today and will continue to do in the near future.

With their theory, Vargo and Lusch have placed design thinking at the core of business strategy, resulting in service design. Naturally, they are not the only ones active in this field, as service design has become very popular in recent years. We see service design as a reaction to coping with the demanding environment that the web and ubiquitous technologies are creating. We are dealing here with a service ecology with multiple users, products and interfaces; where relevancy, time and personalization come together, creating complex networks. On the one hand, service designers always keep an eye on the big picture, making sure that the stakeholders are motivated to be part of a network that exchanges resources. On the other hand, they understand that it is at the human interaction level that people assign meaning to the service.

SDL has 10 foundational premises, which we summarize here:

1. 'Service' is the concept of applying our resources for the benefit of another to exchange value.
2. Sometimes 'service' is indirect because it is provided through complex combinations of goods, money and institutions.
3. Goods (or products) are a distribution mechanism for service provision.
4. Knowledge, skills and the ability to cause a desired change drives competition.
5. We all live in a service economy.
6. Value creation is interactional. The consumer is always a co-creator of value.
7. Companies do not deliver value, they can just deliver value propositions because value exists at the moment of using the service.
8. A service-centred view is customer oriented.
9. The context of value creation is a network.
10. Value is a 'human experience'. It is uniquely determined by the beneficiary in a particular context.

(Based on Vargo and Lusch, 2007)

4.5 FOSTERING SOCIAL CAPITAL

The first time the term 'social capital' was officially coined was in 1916 in an article by L.J. Hanifan regarding local support for rural schools[8]. It referred to the personal investment on the part of the community. Since then, there have been many definitions of social capital. We like the definition used by Dekker and Uslaner that states that social capital is the value of social networks, bonding similar people and bridging between diverse people, with norms of reciprocity[9]. They emphasize that social capital is fundamentally about how people interact with each other and how value is created by those interactions.

"I have a very abstract definition of interaction: an interaction is between two entities, two structures that have a relationship between each other that dynamically changes. So an action by one of the elements causes a reaction by one of the others, that causes in

return a reaction with others and it goes on and on like this. There are relational behaviours which are always dynamic in time, it's never static. Whether it is a technical interaction between two software systems, or whether it is an interaction like communication in human exchange or whether it is between humans and systems. This last type of interaction is what designers are about." — MADDY JANSE

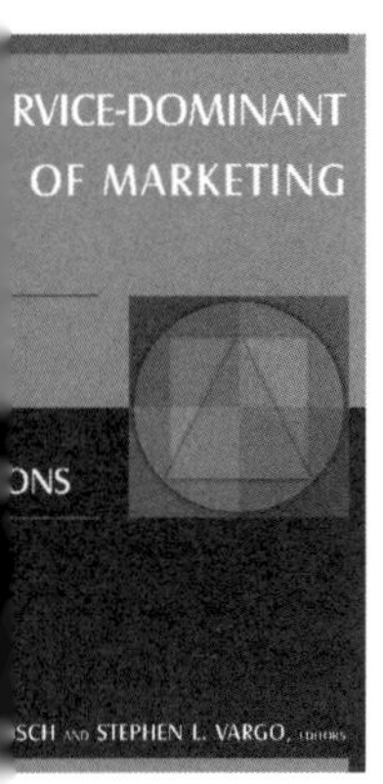

THE SERVICE-DOMINANT LOGIC OF MARKETING, BY ROBERT F. LUSCH AND STEPHEN L. VARGO

So in short, how can the social interactions in Meta Products be valuable? How can a Meta Product foster social capital? Basically we believe it can only happen when people are motivated. As Clay Shirky says "more value can be gotten out of voluntary participation than anyone previously imagined, thanks to improvements in our ability to connect with one another and improvements in our imagination of what is possible from such participation"[10]. There are two ways to lead motivation in a Meta Product that you could consider: 1. Help people build their own goals and 2. Provide the right choices around those goals in order to achieve them. The tricky part will be in matching goals and interactions in a network of multi-users. But we think this is actually what designers can do well if they 'know their materials' (for a more detailed discussion of this, see chapter 5).

Meta Products will result in new interactions between people, via the web and sensing devices that will enable 'conversations'. Julian Bleecker likes to call some of these devices Blogjects (objects that blog): "The most peculiar characteristic of Blogjects is that they participate in the exchange of ideas. Blogjects don't just publish, they circulate conversations"[11]. The new conversations enabled and derived from the web and sensing devices will probably change our behaviour in so many ways that it is worthwhile to reflect upon this, particularly if you are a designer.

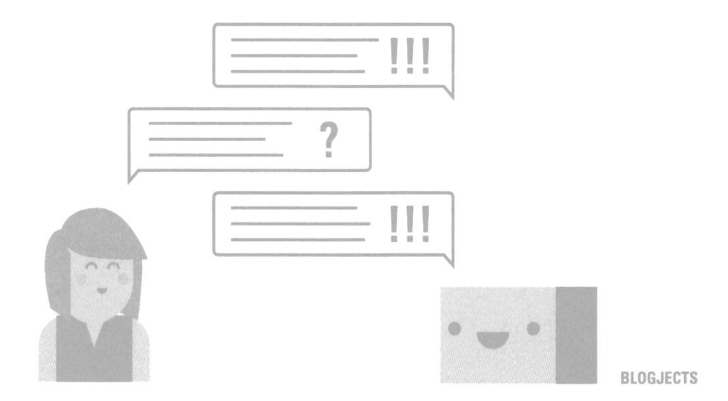

BLOGJECTS

Meta Products will measure or sense data via a device that sends this data to another device or an application on the web to fulfil a service purpose. It can be automated or it might require your input. In any case, Meta Products will enable new interactions and new types of communications between us. What kind of impact will this have? We don't know exactly, but we are sure we can at least design interactions that motivate people to discover their goals and achieve them.

For instance imagine that you are travelling and you do know where to go but you don't like to follow the instructions of a GPS device. You are a more explorative type and you like spontaneity. A Meta Product will know this and might provide you some hints only when a blockage is on the road or when it detects you are really getting lost.

4.6 ENHANCING QUALITY OF LIFE

As a designer you are immersed in deadlines and competition, you want to be appreciated, be able to pay the bills, keep the clients happy, innovate and be successful. And whatever type of designer you may be, you almost certainly have somehow an inner urge to 'do good' to the world around you as well. And this makes sense, because it's not about something like morals, you are simply used to thinking positively. After all, most of your design projects are about improving things and finding solutions. However, that positive thinking and that inner urge is typically not enough to prevent you compromising on your view, and while the client might be happy, you might in fact be polluting the planet or diminishing someone else's quality of life. We believe this is just one of the many consequences of the 'old way' of doing things. We are constantly changing, which means we are also changing our technology to help us actually do some good to the world (or at least stop damaging it).

"We have this world with ridiculous amounts of damage of different kinds and we cannot pretend designers were not implicated in many ways in that. Now we are aware designers are social actors, we are designing the world through our devices, products and services. And that is what I mean with being more in tune with the consequences, to look at the bigger picture and be self-conscious of the impact of what you're doing." — LUCY KIMBELL

In fact, designing for our connected world can help us make people's lives better and enhance their quality of life while running successful business networks as well. More importantly, designing for our connected world can help us be aware of the reach and impact the Meta Products we design can have in the world. This idea is interwoven into the whole mindset of the design approach we are presenting in this book. In fact, much of our inspiration has been derived from research literature on assessing quality of life. First of all, quality of life is defined as an overall general well-being result of the objective life conditions and the subjective evaluations of those conditions based on a set of values[12]. We looked at the way researchers assess quality of life and we saw that this assessment model can also serve as a foundation when designing Meta Products. How? you could use this assessment model to design Meta Products. First of all by empathizing with the way people give importance to what they

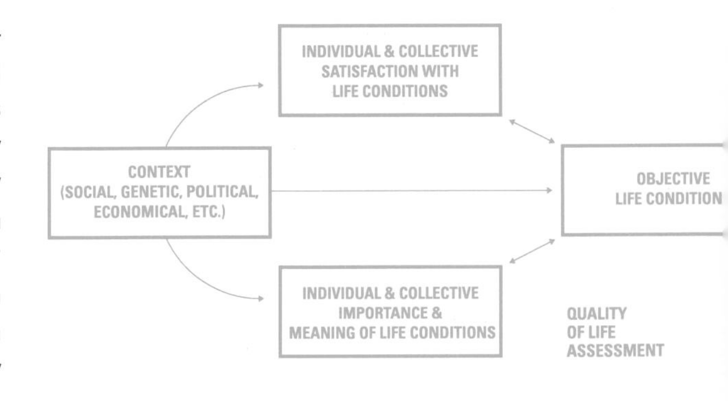

do and how they feel satisfied with their lives, both in collective and individual human dimensions. Secondly, by thinking in 'networks'. When you think in networks, you are able to trace the impact of the interactions you design more easily than when you design with a focus on a brand or on a particular customer. The way Meta Products get to be designed, the way we get to access, share and change information through the web and other ubiquitous technologies will have a big impact on our lives. Will they make our lives better? As Lucy Kimbell expressed, "designing Meta Products is about accountability... who do we want to be accountable to as designers? To other designers? To our clients? This question has to be answered very distinctly."

"I think we live in a contradiction of mindsets. For example, look at how ants behave; they have a totally sustainable economy in one colony. The colony never grows, it is a stable colony and everybody has tasks that are totally defined. If you look at human society, it is similar, although we are not sustainable and that's the key. We, as the ants, have our little clusters, and that's it, we can't handle more than that, nor do we need to. But the other mindset is the technology which enables us to travel everywhere in the world, and access every kind of information, and this is actually infinite. There comes a point when people start to struggle to sort out what they need for themselves, because there is only so much that we actually need. I think we have to focus on that and on the different types of clusters the human society is creating, and the different 'presentation' of the clusters that we have been used to from the old days. For example, in the old days you had a particular kind of contact with the people in the street and your friends from your neighbourhood; now you have virtual contact with people located far away, but even if you have a list of 200 virtual 'friends' you normally only have contact with a few." — MADDY JANSE

Up to this point, this may sound rather philosophical. We are not philosophers or sociologists, but we can get the best out of all these sociological reflections in order to design things that will actually do good to more people in the world.

Tracing the impact on people's quality of life of the Meta Products you design, with a view to fostering improvement, is an ideal that is achievable because the elements of a Meta Product are potentially 'intelligent'. You can see this with the emerging e-healthcare industry. Meta Products enable healthcare services that were impossible before because now we can sense, measure and track our body performance, temperature, blood pressure, sugar levels, and so on. And most importantly, we are finding out ways to filter meaningful information immediately in order to receive appropriate medical treatment. E-healthcare development is one of the most representative industries that reflect the network nature of Meta Products. In the Netherlands, government, technology, device and service providers, specialists and designers are working to develop local e-healthcare networks that can help the aging population to remain at home longer and enhance their quality of life as much as possible[14]. Looking at the increasing aging population this is quite a task for many countries in Europe. But it is clear

For instance, Eli Pariser talked about 'relevancy' during a TED conference in May 2011.[13] There he mentioned how in the quest to find ways to 'filter' all the information now available, Facebook and Google are 'intuitively' editing the information to be 'personalized'. That means that the algorithm of Facebook or Google decides what information you will get, in the name of offering customization and smart ways of filtering information. The intention might be good but actually these practices might block the transparency and openness that the web offered in the first place. Could this take us back to the old model of elites deciding what the rest of the world could or should know? Pariser suggested designing algorithms and interactions that are transparent and that provide people with a means of control. Naturally, providing control and transparency is not the easy way out for the most commercially minded, but perhaps we can actually gain much more as technology helps people to build knowledge in more sophisticated ways, instead of blocking it. Simply put, do you think Facebook and Google have thought about the social impact these personal filters can have? On the one hand, it can create close connections with groups of people all over the world, but could it be that these groups become narrow-minded or isolated as a result?

that the only way to take on this task is to create an e-healthcare network that provides value to everyone involved (elderly, their family, the formal care institutions, the government, and so on). Other industries can also identify possibilities where data and information can be sensed, measured, tracked and translated into useful interactions. For example, achieving efficiency in manufacturing by identifying mistakes earlier in the process, or reducing waste that wasn't able to be measured before, or protecting species that were not accessible before, and so on. In any case, we believe that now is the right time to direct our efforts towards enhancing the quality of life of the people that will use the networks you design instead of adding new problems.

"...the way is to start building it upon the human experience and the human needs. Then you will realize that there are certain things that can leave some room for technologies or new products and services. The value is to reduce complexity...there is no actual need to have new devices or objects that compete with each other." — FEDERICO CASALEGNO

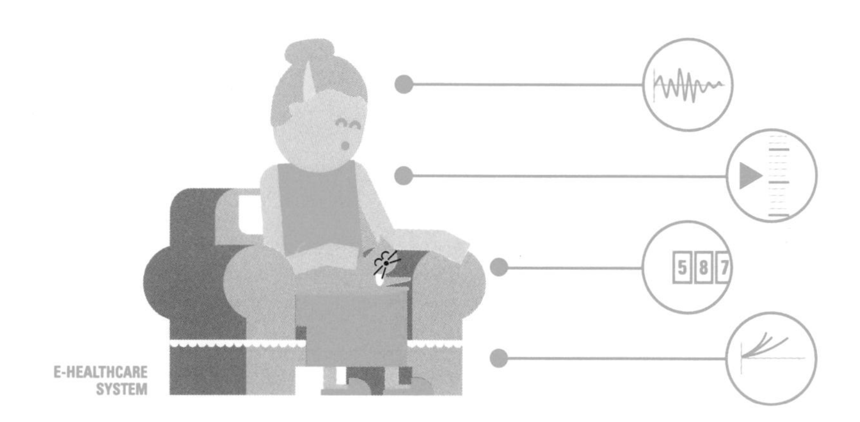

Encouraging meaningful experiences, building networks of value, mastering transdisciplinarity, fostering social capital and enhancing quality of life are our top 5 ideals when designing Meta Products. As you can see, they are thought from a rather holistic perspective rather than an operational one. For this chapter, an operational perspective would be insufficient to convey what we believe are the core guidelines for designing Meta Products. In chapter 5 you will find a more operational perspective around the subject. Rather than being mathematical, chapter 5 will recommend you some techniques. We will also introduce our design approach (Network Focused Design) that aims to guide, inspire and invite you to construct the best practices for your design activity.

CHAPTER 4 — SUMMARY INSIGHTS

1 Learning means modelling your own world-view. It involves your own prior knowledge, the present experience, and the situation itself. This is how we learn, or in other words, this is the way we assign meanings and build knowledge.

2 'Encouraging meaningful experiences' is an ideal placed at the centre when designing a Meta Product (the people or the users are at the centre of the design process), the way these experiences are orchestrated, or in other words, the way to make them happen, is by building networks of value. This means creating ways to motivate the members of a network to exchange their resources. During this process, new members, new resources and new methods of exchange might arise. The information stream would be 'the fuel' of it all.

3 Motivation is the driving force through which we achieve our aspirations and it is closely related to the way we assign meaning to everything we do.

4 Building a network of value means providing a service that is meaningful to the people in a network and then offering them a platform to adapt and grow that service as they need it.

5 Transdisciplinarity means that disciplines working together cross their boundaries with the aim of finding new solutions. It's all about interventions between disciplines that are necessary to solve problems that have not been — or cannot be — solved through conventional multidisciplinary practices, or when the aim is to innovate.

6 Transdisciplinarity is particularly important if the network is complex in terms of feeding or analysing dense streams of data, multi-users, multi-interfaces and parallel activities, through spaces and devices — in real time.

7 Social capital is fundamentally about how people interact with each other and how value is created by those interactions. Meta Products are potential ambassadors of social capital.

8 Social capital is possible when people are motivated: There are two ways to lead motivation in a Meta Product that you could consider: 1. Help people build their own goals and 2. Provide the right choices around those goals in order to achieve them.

9 Assessing the quality of life of people when designing Meta Products means: 1. Empathizing with the way people give importance to what they do and how they feel satisfied with their lives, both in collective and individual human dimensions. 2. Thinking in 'networks' by tracing the impact of the interactions you design.

AIRSTRIP PATIENT MONITORING

AirStrip Patient Monitoring lets healthcare professionals remotely access real-time patient data directly from the hospital and send relevant information to a handheld mobile device. It is meant to be used in hospital environments where doctors need access to critical patient data or regular bedside monitoring data whenever necessary. Their service consists of three key parts: a measuring interface, a web server and a mobile device.

The measuring interface is connected to existing hospital monitoring devices. It reads out and then sends patient data like blood pressure, heart rate and temperature securely over the internet to a web server. The clinician can read this patient data on a mobile device via an app that communicates with the web server.

AirStrip Patient Monitoring offers clinicians a valuable service in that it bundles and presents relevant information of patients in a single app. They can view data trends and receive alerts if something goes wrong, saving plenty of time by only having to launch a single app.

With hundreds of hospitals using AirStrip already, AirStrip Technologies have an experienced point of view on how to achieve what they call "meaningful mobility" in healthcare. Meaningful mobility is defined by AirStrip Technologies as "native mobile technology that improves clinical decision making at the point of care through data transformation and the secure, real time, and ubiquitous delivery of visually compell ingintelligence by incorporating evidence-based medicine and knowledge-based prompts."

AirStrip's definition of meaningful mobility seems to be in tune with our approach towards Meta Products. We strongly believe that products shouldn't be connected just because they can. It all starts with understanding how people build their aspirations and how they perceive the value of new interactions. In this case, the value is clear. Health specialists have immediate access to patient data, without having to meet each patient or scroll through huge numbers of documents. This not only saves them time, but makes their work less error prone, resulting in improvements to the quality of the care service. By focusing on the meaning & value, instead of on mere technological possibilities, Airstrip came up with a product that transforms the way patients are treated.

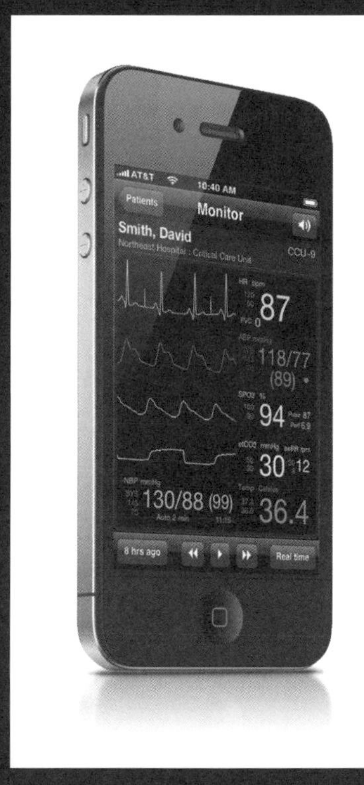

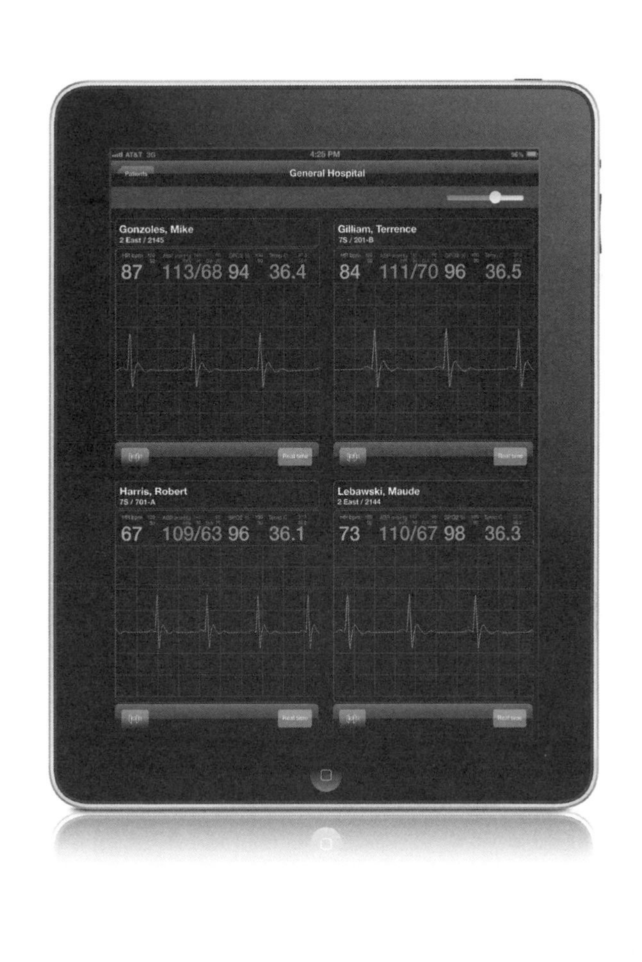

IMAGES PROVIDED COURTESY OF AIRSTRIP TECHNOLOGIES, INC.

FEDEX SENSEAWARE

SenseAware is a small device from FedEx that measures in-transit conditions. It's mainly used for shipments that are vital or critical. The idea is that you put the device inside your package, where it will perform its monitoring during transportation. SenseAware has a couple of sensors that measure if a package has been opened or exposed to light, its location and its temperature along the way. You can also set threshold presets that notify you immediately when these are reached. It was launched in 2010 mainly for the healthcare and life sciences industry, where it is used for monitoring transport of pharmaceuticals, human skin tissue, organs and medical equipment. The security delivered by SenseAware for these types of shipments is usually worth the extra costs.

Using a web service, you can track your shipment statistics in real time. By monitoring the actual location of your shipments, you can take timely actions and coordinate quickly with supply chain partners. It's possible to set up a geo-fence within which the shipment may be transported, raising an alarm when it exits the supplied perimeter. You can also invite your customer to see the actual statistics of your shipment by sharing them via the SenseAware web platform.

Like the first shoe sensors of Nike+, SenseAware gives existing goods an extra data layer that provides more information. This information shadow can then be used to analyze and monitor the package in real time. In other words, the package acquires more attributes, which you can autonomously analyze with a computer. For the transportation world, this is great added value. Now it's possible to know what happens to a package during transit. Instead of waiting for an inspection on arrival, you can now take actions immediately. When transporting vital goods, like skin tissue or organ material, this literally can be a life saver. SenseAware is also a great tool for optimizing the supply chain as a whole. For large transport businesses, SenseAware can be used to track which conditions have an effect on the shipment, and act accordingly.

SenseAware aims to make the supply chain processes more efficient. It does so by understanding how an information-fuelled network can bring value to the logistics processes. It considers both the human dimension in the design of the touchpoints or interfaces that communicate with people, and the functional complexity of the communication between objects in the overall network.

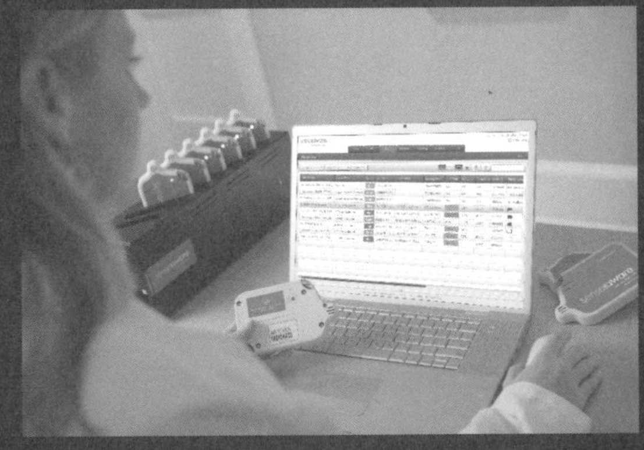

IMAGES PROVIDED COURTESY OF FEDEX CORPORATE SERVICES, INC.

Which kind of process do you think is required to design for our connected world?

MADDY JANSE
Philips Research

Designing a system or creating a product at the front-end is being done a lot, but only around 2% gets really to the end. One problem at this moment is that you can get access to certain elements of the system, for instance of the smart home, like lighting control, security elements, and so on. They are all there, but they don't talk to each other, they're islands. So you first need industries to understand what the standardization and open-source software can mean, before they can really see the economic advantage and make the process change.

What kind of value do you see in the Internet of Things?

JAN BUIJS
Delft University of Technology

With Meta Products and internet in general the systems are more dynamic while new uses of time, personalization and location arise. We can see it already, for instance, the interface of my smartphone is different than yours because I can customize it to the services I use, all in one place. The same will happen with technologies converging into cars for instance, making the car a big technology box of services.

ANNA VALTONEN
Umeå University

The internet itself is just a means in the process. It is actually what we do with it that counts. The distinction between online or offline is not what counts. In some situations people want to feel that they are in charge and take conscious choices. So if they know what the device can do for them and what they can do with it, it is irrelevant if it is online or offline for them. Indeed some functions have to be transparent nowadays precisely because people want to feel they are in charge, and for that you have to give the choice to people whether they want to interact with the device or the function or not.

FRIDO SMULDERS
Delft University of Technology

Apart from the Internet of Things, our methodologies are not good enough yet to achieve true collaboration. We have been very much focused on the content and the tangible elements of designing and we completely overlooked the intangible elements. And with this I also refer to the knowledge-sharing and the collaboration dimensions of designing. My research is focused on making these dimensions explicit. There is a difference between designers in the first place, and this knowledge remains somewhat hidden because we don't teach how to collaborate. Why does it take 15 or 20 years to become a very good designer? It means that there is an element involved in practice we can't teach (yet). And collaboration is in my opinion that very practice-related element!

ANNE LISE KJAER
Kjaer Global

By putting the people-centric dimension back into the centre of the design process, we can identify the true needs of people in a connected holistic vision — only then can we create meaningful products and services. Tomorrow's designers need to embrace whole brain-thinking to decode the complexity — this is the only way to navigate complexity.

ANNE LISE KJAER
Kjaer Global

The key is to predict where the real value is, and who is willing to pay for it. It is not enough if you are the only one perceiving the value. We must remember that our businesses will become part of an information system. This information system becomes an added asset that has the ability to capture, compute, communicate, and collaborate around the stored information. These 'smart' assets can make processes more efficient, give products new capabilities, and spark novel business models.

ALEXANDRA DESCHAMPS-SONSINO
Designswarm

Value is a complex concept that has a multitude of faces, but perhaps in the overall scene of the Internet of Things there is value in the possibilities that it gives us to measure, manage and be informed whether what we are doing is efficient or not. Especially when the effects are on larger scales, such as the consumption of energy, for example. So when this value is recognized it can actually change behaviours. Another similar value but in an individual perspective is 'personal informatics', the Internet of Things offers a possibility not only to measure and track my personal information, but to act in intelligent ways. So for example how much I run, how much sugar there is in my blood, how much energy I consume. It gives us intelligent tools to make good decisions about ourselves.

JENNY DE BOER
TNO

It depends, value is identified when the users and stakeholders get to experience the product or service, that's why we co-design, particularly to create value together so that they can recognize it. So there is no one answer. Actually, we believe that a product that is co-designed will be adopted more successfully than one that wasn't, because they will see the benefits or the value more clearly.

CHAPTER 5 — *Designing Meta Products*

Network Focused Design

PHASE 1 Visualizing networks

PHASE 2 Setting a direction

PHASE 3 Designing the Meta Product

creative practice / reflection / intuition / design thinking / cognitive styles / network mindset / network layers / iterative process / new ecologies / system vs. network / empathy by reflection / observational research / co-designing / re-enacting / role-playing / personas / journeys / framing aspirations / abductive thinking / exchanging resources / change / motivation / opportunities / creative facilitation / new values / idea generation / motivated stakeholders / design guidelines / revenue flows / value exchange / new network / information-fuelled network / designing interactions / perception / engagement / emotions / aspirations / quick testing

5.1 THE GOLDEN RULE

Good carpenters know their wood from A to Z. They know what type of wood it is and how it behaves in different circumstances. They know how it reacts to the environment, to weight, to friction and over time. They have a clear idea whether your client's design can be produced using wood or not. They know the limits to which they can force the nature of the material, and they respect those limits.

The same has been true for quite some time. For instance, eighteenth century carpenters knew what they could transform, so they knew the different types of wood and their characteristics. They knew how to transform the wood, so they knew the tools and the processes. And most importantly, they knew why they transformed the wood: so they knew beforehand whether they were making a very ornate chair for the governor, or a less fancy chair for their old neighbours. So, by being aware of these three things they knew what material was appropriate to transform it. These are basic activities we've learned to do over time. We've learned them because we've understood that the very basic activities of creation have always followed one golden rule: know your materials.

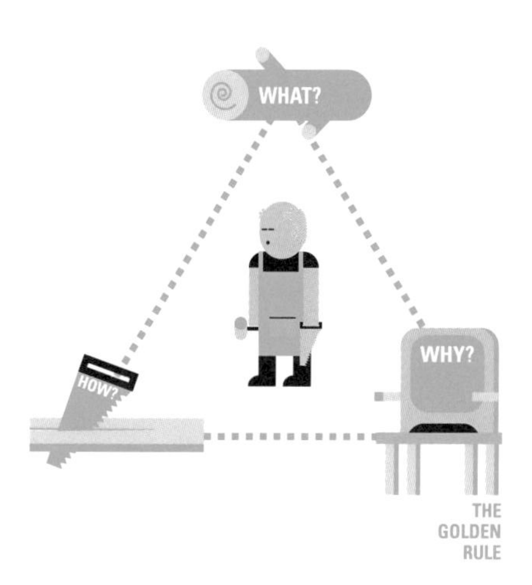

THE GOLDEN RULE

This one rule hasn't really changed over the centuries. At least not fundamentally. The only difference over time is that more and more elements have come into existence. But too many elements can make the whole overwhelming, which is why you've learned to analyze them, categorize them and synthesize all possible answers. You've learned methods and techniques to do so; and now the design world is flooded with terms such as design thinking, creative facilitation, team work, co-participation, co-creation, and customer experience. In this chapter we will be recommending some of these methods, tools and techniques. Each of them has been part of our learning process throughout human history, helping us understand the answers to the golden rule, and they are not just limited to design practices.

Today, we're going through an important moment in history where the questions we need to answer in order to follow the golden rule are constantly changing, and we're being forced to reflect in order to make good decisions.

In this book we invite you to join us in that reflection process, after all, "reflecting is a normal human activity that is essential for development and allows personal growth by learning from our experiences, mistakes and successes", according to Nakielski[1]. It's no coincidence that many theorists have dedicated their work to understanding reflection. In fact, we were very much inspired by the theories of Reflection and Action of Schön[2] in which he describes the reflective practitioner as someone who learns by reflecting on current experience and applies that learning to future practice. The reflective practice is an unstructured self-regulated approach that aims to help understanding and learning. It is most commonly used in the health and 'helping' professions, but we find it very similar to any design process. Except that designers also rely on their intuition to innovate. Or how Charles Sanders Peirce calls it: 'abduction reasoning',which basically means making educated guesses when there is no one single possibility or 'the logic of what might be'.

We are not the first or the only ones inspired by these great theories. In fact, we think that disciplines such as human-computer interaction, user-centred design, interaction design, user experience, service design, and basically any process that claims to be 'iterative' and 'holistic' have their roots in the reflective practice and experimental learning theories that arose in the late 70's. We find it refreshing to look at Rolfe's reflective model[3] which consists of answering three questions: What? So What? and Now What?; describing, interpreting and imagining the future are the key activities of a reflective practitioner. So they are also key for anyone who designs.

This brings us back to our golden rule: know your materials. How can we know them? First of all you have to understand that 'knowing your materials' is a reflective process to discover what can be designed, why it can be designed and how it can be designed. These are three basic reflective questions to ask in order to follow the golden rule, and this is our starting point when designing Meta Products.

PAPER PROTOTYPING WEB INTERFACES

What happens in your mind when you follow the golden rule? Understanding the mind of designers has been an intriguing subject for quite some time. Back in the 60's, the design activity started to gain the attention of cognition specialists and began to be perceived as a high-level mental process that allows us to create and change our environment in order to improve it. Yes, design plays an important role in our society, and although we might not save the world through our designs, we can try to make it a better one. And let's not allow terms to confuse us. We're referring here to 'design' as the creative and reflective practices that have existed in many professions for very many years, and that have been responsible for shaping and reshaping our world. It's the 'design' activity we've been defining and learning to master over time. So it's hardly any wonder that nowadays we're hearing so much about 'design thinking' not only in strict design fields but also in business and other areas where innovation and change is necessary.

MOSAIC THINKING

"...twenty years before I heard everyone talking about 'design thinking' I called it 'mosaic thinking', which is a bit similar I guess or maybe it means the same I don't know... but what I mean with mosaic thinking is that you have to concentrate on the little stones, understand them and position them in the right place, it's maybe a matter of millimetres... then you have to zoom out to see if the whole picture is right, and only then can you see whether you have to move it a little bit to the right or to the left, then you zoom in again and make the changes... and so on. This continual changing from the abstraction level of thinking and acting is what I call mosaic thinking. And all good designers do that." — FRIDO SMULDERS

Donald Norman (consultant of the Nielsen Norman group and columnist of the design website Core77) explains that design thinking means stepping back from the immediate issue and taking a broader look. It requires thinking about systems:

realising that any problem is part of a larger whole, and that the solution is likely to require an understanding of the entire system.

This is indeed the typical way of viewing a reflective practitioner: holistic most of the time, analytical at other times; intuitive most of the time and systematic at other times, all in order to find and frame the problem together with its solution. It's not just about generating solutions but also about interpreting the problem. Defining the criteria for evaluating solutions is also a task for the reflective practitioner. All this happens in loops or iterations. In the 80's and 90's, literature referred to the activity of design as a mental process that involves dealing with open-ended and 'ill-structured' problems where there is never a single optimal solution. What differentiates a good designer or good design team from a bad one is the insight and the ability to perform lateral transformations and set-shifts when dealing with open-ended and ill-structured problems.

A way to adopt the mindset we are talking about is perhaps to start reflecting on your preferred way of thinking. In academia these are referred to as 'cognitive styles'. Academics acknowledge that your cognitive style might affect the way you learn, solve problems or perform tasks[4]. Design tasks are open-ended and ill-structured, so in theory, your cognitive style should match the nature of the design tasks so that you can perform them in an effective way. Basically there are four main preferred ways of thinking: either holistically, analytically, as a verbalizer or as a visualizer[5]. If you tend to think in a holistic way it means that you would naturally perform well when several equally acceptable answers are required to be generated. You perceive information in 'one big picture' and approach it in an exploratory way. You tend not to pay attention to minor tasks in order not to lose focus, therefore you're good at scanning large amounts of data and recognizing patterns. On the other hand, if you tend to think in an analytical way you would naturally perform well when just one correct answer is required. You would do this quickly and very precisely, especially when examining less amounts of data.

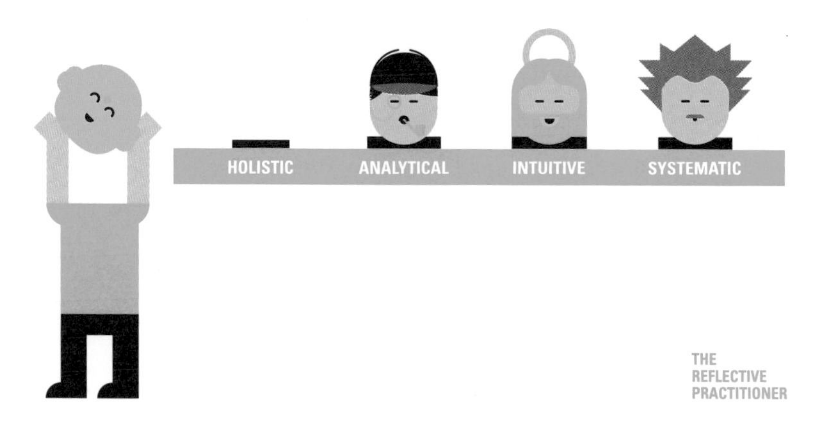

THE
REFLECTIVE
PRACTITIONER

You'd be good in completing tasks step by step and you'd feel more comfortable when working individually.

If you tend to think in words you'd be what theorists call a verbalizer. You might prefer to get solutions and find inspiration from abstractions. Stories might be your way of imagining the future and you'd be good at writing them down. On the other hand, if you tend to be more concrete and use images and drawings to find solutions and to communicate your thoughts, you're what is called a visualizer.

Throughout your life experiences, you learn techniques to adapt your mind style depending on the situation, for example when you have to solve a maths problem, write an essay or ask someone a favour. So, take a few minutes to reflect about when you feel more comfortable when trying to find a solution or when you're learning something new. There's no right or wrong, it will simply help you to identify your design role and to understand other roles in your team. This is important to think about, particularly before starting any creative activity.

The conclusion could be then: get to know your materials as a reflective practitioner. In other words, discover what, why and how to design something through a process of reflection where you may have to adapt your mindset, from holistic to analytical to verbalizer or visualizer if required.

We believe that designing Meta Products demands new levels of mastery of the golden rule and design thinking from designers. Because with the web and ubiquitous technologies we create streams of information feeding new interactions and connecting more people, things and spaces in new ways. Information and connectedness are the new materials of designers, as these are 'the fuel' of the interactions we design. Designing a Meta Product can involve considering multiple users, multiple services, and multiple products at intermittent times, some elements are visible while others are invisible, some elements are human-initiated while others are automatically generated. Hence, designing Meta Products requires a network mindset, transdisciplinarity and great communication skills from team members involved in a project. In chapter 3 we explained what we think are the main challenges Meta Products pose to designers today, which makes it all more exciting as well: network, meaning, process and platform.

"Is designing an iPhone more complex than designing a shoe? Why would Meta Products be more complex than what designers are designing now? The multiple interactions? The artificial intelligence? The multiple components? A car or a plane has a multitude of components and these are not Meta Products and they've existed for many years, and still designers are busy with them. I have some difficulty then defining 'complex'. Designing is always challenging, and it's always about the future. In principle, there are trends and things happening that designers can pay attention to in order to design products and services. With the Internet of Things you can play with the network layers and boundaries. Every added layer creates more dynamism but designers have to focus on what is actually valuable and useful."

— JAN BUIJS

Bearing in mind the golden rule, the design mindset and the Meta Product challenges, we tried to come up with an approach that could guide us to design Meta Products. We named it Network Focused Design.

"...the implications of Meta Products for designers are (a) making sure they work with sociologists and anthropologists who have useful concepts, tools and approaches for trying to understand and describe what people are doing in their engagements with such products, services and spaces, and what this might mean at collective rather than individual levels; and (b) having iterative design processes that allow this developing and emerging understanding of what things might mean to continue to shape design, testing, development and delivery." — LUCY KIMBELL

5.2 NETWORK FOCUSED DESIGN

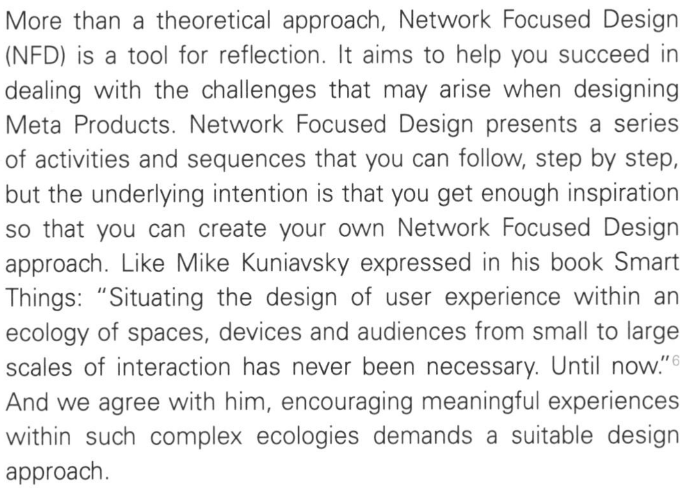

More than a theoretical approach, Network Focused Design (NFD) is a tool for reflection. It aims to help you succeed in dealing with the challenges that may arise when designing Meta Products. Network Focused Design presents a series of activities and sequences that you can follow, step by step, but the underlying intention is that you get enough inspiration so that you can create your own Network Focused Design approach. Like Mike Kuniavsky expressed in his book Smart Things: "Situating the design of user experience within an ecology of spaces, devices and audiences from small to large scales of interaction has never been necessary. Until now."[6] And we agree with him, encouraging meaningful experiences within such complex ecologies demands a suitable design approach.

SMART THINGS, UBIQUITOUS COMPUTING USER EXPERIENCE DESIGN, BY MIKE KUNIAVSKY

With Network Focused Design we combine the most relevant mindsets of practitioners and theorists involved in the field of Meta Products. These practitioners are ethnographers, service designers, user experience- or human-computer interaction adepts, interaction designers and ubiquitous technologies researchers and experts. We looked at their ways of doing things and tried to find out what a transdisciplinary approach would look like in order to get the best out of all of them. We mingled methods, techniques, theories and processes and tried our best to simplify them into 9 hands-on steps. NFD is an approach that will help you to identify opportunities where new relationships of value can be created. This approach focuses greatly on taking you out of the technology world for a while before actually designing information flows and interactions. It also guides you to think in networks and to focus on the relationships that built them. This has three main aims: the first one is to help you design meaningful Meta Products that truly improve people's lives. The second one is to help you find innovative opportunities in your industry or field of interest. And the third one is to find interesting business cooperations between the stakeholders.

Our world is formed by networks of products, services, spaces and people. Networks have a purpose according to the relationships they enable. So, as you might have guessed right now, as a designer you transform networks.

We like to use the term 'network' because it refers to an 'open system'. The dictionary definition of 'system' is a group of interacting, interrelated, or interdependent elements forming a complex whole. However, though a network seems similar, it reveals a more 'open' nature; it is something resembling an openwork fabric or structure. It makes us think of flexibility and growth. We're used to buying closed systems, let's say a refrigerator. You buy it complete with doors, shelves, and the mechanisms inside and you plug it into the power supply. You can't make it work better or add functionalities to it or turn it into something else because when you bought it, it was already finished. The production process was closed and it was during this process that the definition was formulated as to how and which elements should work together to make a refrigerator a refrigerator and not something else. If one of the elements fail or is absent, it will affect the whole system or will render it absolutely useless. A network on the other hand is far more flexible and it's possible for its elements to re-invent themselves. A classic example of a network is our road network. We built it to connect towns in order to make it easier to transport goods and communicate with each other. The more connections in a road network, the more activities are enabled, hence economy is enhanced. Another example of a network is the telephone. It consists of telephone lines, fibre optic cables, microwave transmission links, cellular networks, communications satellites, and undersea telephone cables all seamlessly interconnected through switching centres which allow any telephone in the world to communicate with any other. In both examples, parallel activities occur and it is possible for different events to occur at intermittent times. In both these networks, it's always possible to take more than one route. Actually, the more people own telephones, the more valuable the telephone is to each owner and the better the service becomes.

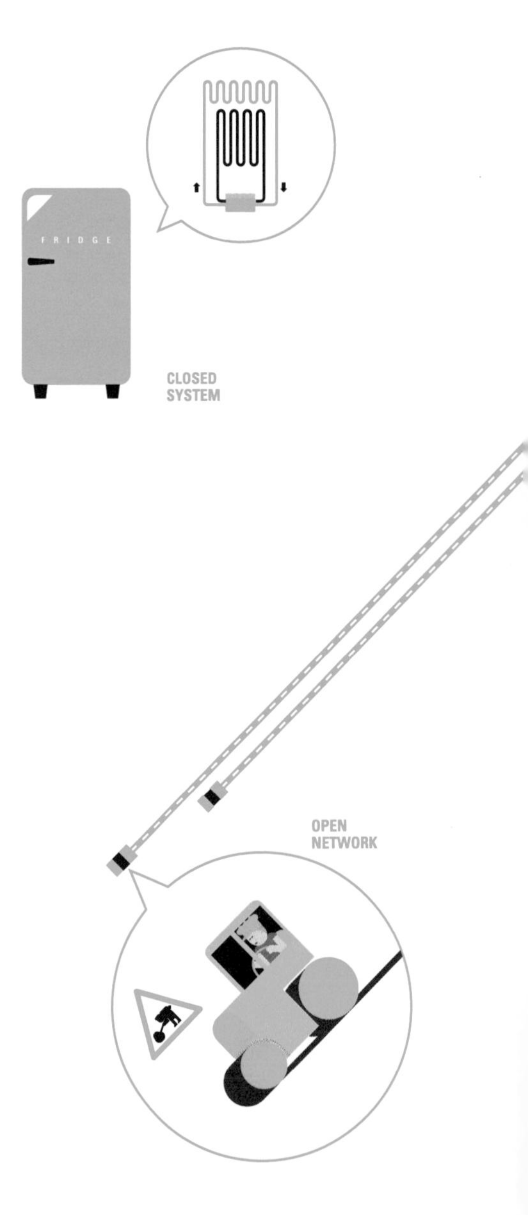

Although systems and networks may sound similar in their dictionary definitions, what we are trying to say is that designing systems is different from designing networks. The difference lies in their nature and in what they trigger in you. A system tends to be closed and a network tends to be open. A system might require a more analytical way of thinking and a network a more holistic reflecting process. Designing Meta Products means you will often deal with the network nature and a bit less often with the system nature.

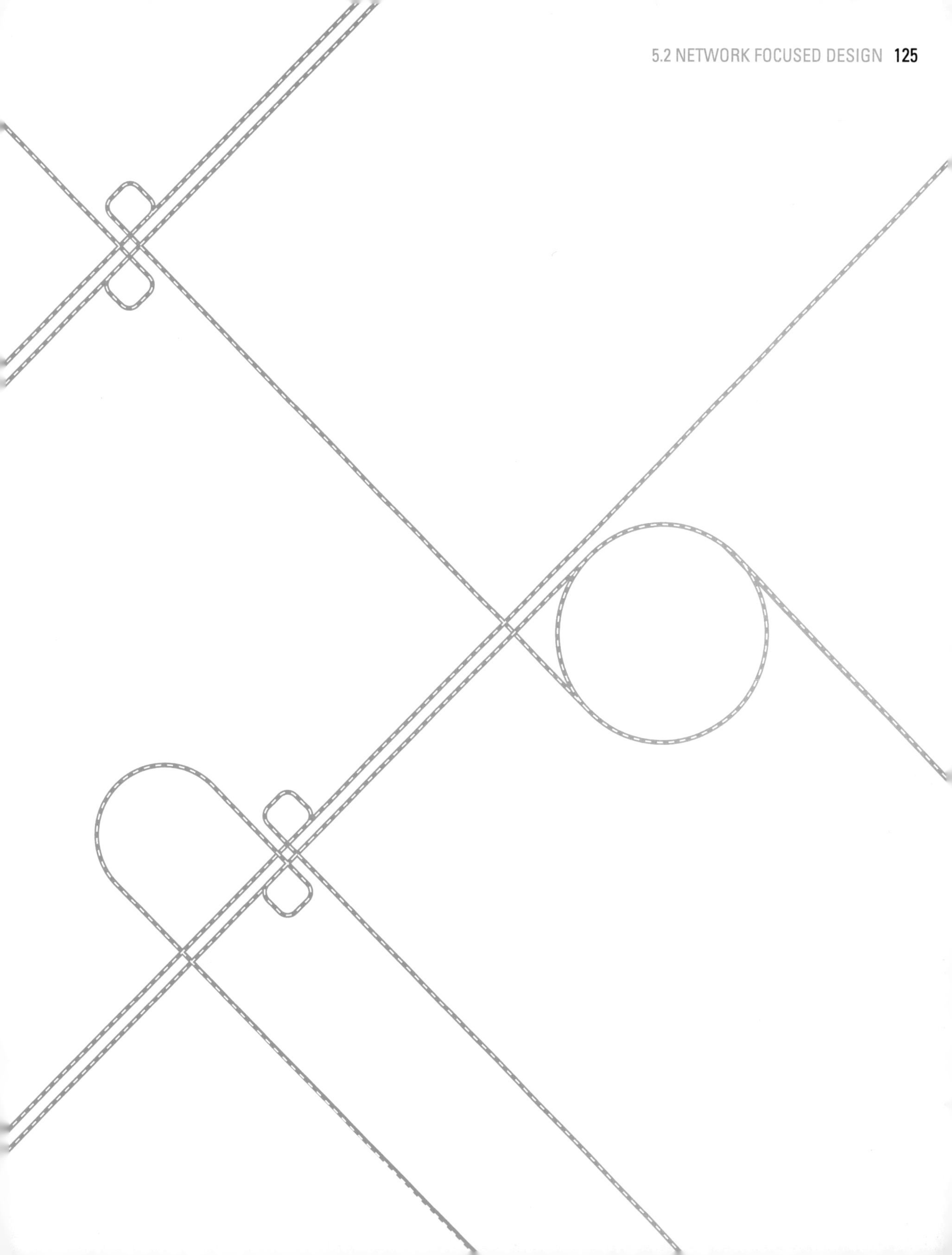

NOTE: YOU CAN DOWNLOAD THE NFD APPROACH ON WWW.METAPRODUCTS.NL

PHASE 1	**A** EXPLORE THE CONTEXT IN ACTIONS	**B** IMAGINE HOW RELATED ACTIONS HAPPEN IN TIME	**C** IDENTIFY ALL OTHER 'ACTORS' AND SHOW THEIR RELATION-SHIPS
PHASE 2	**D** IDENTIFY CHALLENGING RELATIONSHIPS	**E** REFRAME THE NETWORK	**F** DEFINE YOUR CRITERIA & VISUALIZE THE NEW NETWORK
PHASE 3	**G** IDENTIFY THE COMMUNICATION FLOW	**H** DESIGN THE TOUCHPOINTS	**I** TEST

Network Focused Design

A Explore the context in actions
B Imagine how related actions happen in time
C Identify all other 'actors' and show their relationships
D Identify challenging relationships
E Reframe the network
F Define your criteria & visualize the new network
G Identify the communication flow
H Design the touchpoints
I Test

PHASE 1 VISUALIZING THE NETWORK

The only way we can begin a design process is by learning to challenge our own thoughts. One way to learn to do this is to explore and reflect. It's very helpful to do it in a visual way because when you encounter a context to explore, it's easy to become overwhelmed by the mass of information on all sorts of levels. By visualizing, you innately try to force yourself to link high-level information (such as abstract keywords) to low-level information (such as drawings or a scheme). Visualizing can simplify the process of analysing open-ended and ill-structured situations, and only then can you avoid an overload of information and be able to really see the big picture. Moreover, while visualizing a network, you will inevitably think of the relationships between the elements that constitute it. And this may prompt you later to challenge your own thoughts and find new relationships or new elements.

The first three steps of Network Focused Design are all about describing actions and revealing where the relationships of value lie. Why people are willing or not willing to exchange value is a key reflection at this point. This is when you will start making sense of the situation. Sense-making can be defined as "motivated continuous effort to understand connections in order to anticipate their trajectories..."[7] In these first three steps we suggest plotting the connections in a way that helps your sense-making process.

MAIN ACTOR +  MAIN ACTION

EXPLORE THE CONTEXT IN ACTIONS

This is the first and most important step of Network Focused Design. Looking at and assimilating our world is something we do every day. However, in that process, we have all built certain representations in our mind of how the world is constructed. This could stop you from seeing other realities or it could prevent you from imagining different relationships that might work better for other people. Fortunately, there are some tools and techniques available to help us confront our ideas and see the world with fresh eyes. Moreover, there are also ways to help us deal with the large amount of unstructured information that we could get when exploring the context. We recommend the tools that basically aim at achieving empathy by reflection.

"...the way is to start from the human experience and human needs and then you will realize that there are certain things that can leave some room for developing technologies and new products and services. What we need right now is to reduce complexity: there is no need to have new devices or objects that compete with each other..." — FEDERICO CASALEGNO

For example, when you start a new project, or when you're looking for innovative solutions, try to identify the most important actor and the main action. It doesn't matter if you don't really know, remember we're just framing the context while exploring it. So, just pick what you or your team think the main actor and action are.

Now that you have chosen the main actor and action within the context, try to confront the mental model you have built and achieve empathy by observing, describing and interpreting the

Example
Innoviting is a young company dedicated to finding meaningful ways to prolong the quality of life of elderly people that live alone at home but require special care. The first on-going project is called Livind: Assisted Daily Living. Innoviting explored the context by observing the behaviour of the elderly in their own homes, and tried to identify patterns that could provide design directions. What they noticed was that the elderly differ enormously among themselves. This made them realize that they had to shift from a technical-driven perspective to a human perspective. With a technical-driven perspective it was simply impossible to standardise behaviour patterns and it would never render a good solution. Whereas from a human perspective, the patterns become broader and multifaceted. From this perspective, Innoviting started to observe the relationship between the caregivers and the elderly. They held interviews and casual conversations and got immersed in the daily routines and in the emotions of both actors. They noticed that many of their assumptions needed to be re-assessed. Tim van den Dool, the initiator of Livind said: "99% of the time when we were exploring the context, I was really biased because I just wanted to make a certain technology fit while the people weren't really ready. Later on I realized we had to focus on the needs of a specific group with a specific network around it." After re-evaluating their assumptions through observation and reflection, Innoviting saw that the most important actor at the moment is the caregiver and his/her main action is to provide specific and frequent attention to an elderly person with Alzheimer's disease.

context of the main actor. This means going where the main action is happening, and observing what the actor does.

Recommended techniques for observing: observing is a reflective practice that can give you great insights for understanding a problem or a situation, or that can generate new unexpected ideas and associations. There are, however, some basics to learn before actually getting out there and observing. First, define the aim of your observation. Define whether you will observe with the aim of describing a situation, or describing and 'inferring'; or describing, inferring and evaluating. If you want to learn about a situation that's unfamiliar to you, it will probably help you to start being descriptive. If your aim is to be descriptive, you have to be careful to try to write down (or video-tape or audio record) everything about the situation you can observe and not only what you might think is relevant. So try to put your assumptions to one side and just use the material to describe a situation. On the other hand, if you are already more or less familiar with a situation, we would recommend using inferential observations. This means that you will observe and describe, making inferences about what is observed and the underlying emotions. For example, you might observe a girl banging a sweet vending machine. From this observation you infer or assume that she is frustrated by the way the machine works. Finally, if you are already familiar with a situation but you have a specific question about it, you may observe in order to describe, infer and evaluate. This means that you will make judgements based on the behaviour. For example you may question whether sweet vending machines provide an 'enjoyable' experience to children. 'Enjoyable' is already an evaluative judgement.

Once you've defined the basics for the observations you intend to carry out, you have to define the mode of observing. Is it going to be obtrusive or unobtrusive? Obtrusive means the subjects being observed are aware of the research. You could then hold informal interviews, or just observe what they are doing directly, or actively participate in the life of a group or in a situation, plan collective discussions, or ask for the narrative of life-histories of people. You could also choose for a mix of tools. For example you could use direct observation for a short period of time and then conduct a 'cultural probes' or 'diary' study[8]. With cultural probes you provide a diary booklet to people in the context you are exploring that will help them share their

CULTURAL PROBES

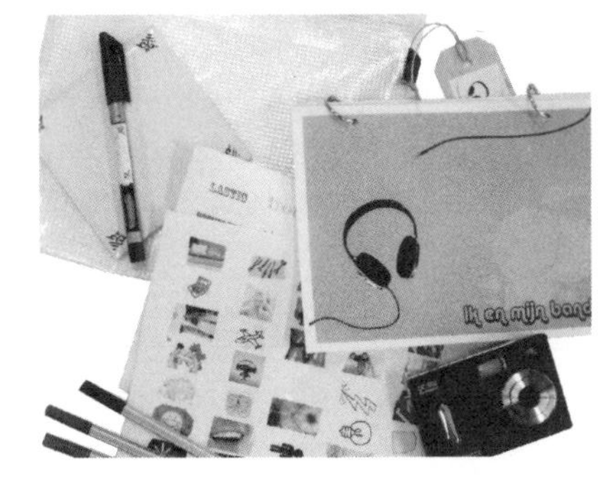

experiences. The design of the booklet and the extra elements you could provide will depend on the individuals and their abilities to express themselves. For example some people express themselves better if they record their own voice, others like to write, others like to use stickers. There's no one right solution and you will have to use your intuition to choose the best form of expression. Then, they will track and input their experiences autonomously. The whole process might last a week or two.

Try to focus on what they do and feel, and not so much on what they say. Getting insights from people is very good, you could also co-design with them. However, as Jenny de Boer mentioned during an interview: "a lot of users cannot co-design because they're not used to it. Co-designing has the challenge to get the community active to create a product they get engaged with and use".

By exploring the context with empathy and in a reflective way, you will understand the relationships between the people, products, services, spaces and other resources, and you might get to find a sweet spot where new relationships can be created and where new sources of information can lead to new interactions.

If you have a tight schedule for this step, you could also try to re-enact and role-play. You can do this anywhere, and with very simple materials.

Recommended techniques to re-enact and role-play: most of the observational research tools recommended above generally require you to reserve adequate time, depending on the objectives of your observation. But if you're a small design agency you can't afford to spend too much time in observation exercises. If that's the case, an alternative could be to re-enact and role-play. There are no strict rules about this. The only point of this is to try to make it as close to reality as possible by using cheap props and making efficient use of your time. For example, get your team together and assign one person the protagonist role (main actor). This could be your client or the main user of your service. Assign auxiliary egos (related actors) to other members of your team who will take the role of people around the protagonist. Then assign other members to be the audience that will react to the play. You could take the role of the director and, finally, prepare the space you're located in with props that would be present in the context. This can be done very easily, for example by using masking tape of different colours to delineate spaces and creating the different scenes. Cardboard boxes for products and hanging fabric for walls are also useful. Here, if you have a verbalizer style you will be able to make very good use of it because it will be very handy to use narratives and write dialogues that could prompt non-verbal gestures to make the role-play more insightful. If you don't have the time to create dialogues you could also use 'gibberish', in other words make a pseudo-language like 'dada-dabla-bla' instead of words. Your facial expressions and exaggerated intonations and movements would be key to your own learning of a situation and that of the team. The most important element when re-enacting or role-playing is to have an open attitude and positive team dynamics.

During the re-enacting and role-playing processes, you will notice that a lot of questions will arise. This is because in order to recreate what happens in real life you will have to be as descriptive as possible. It's very important to mention here that reflecting and observing is not meant to produce quantifiable results. For your own understanding only, you could make use of categorization techniques, lists, and other ways to cluster the data you find, and to communicate the findings to others in some cases.

PERSONAS

One way to help you narrow down your findings and to share them with others is to build 'personas'. Personas are archetypes of the main actor or actors, and although they are fictitious, they should preferably be created based on real information so that you can depict their goals, motivations, expectations and aspirations within a certain context.

It's possible that in the course of this exploration you will get several personas that are very close to the main one (the main actor). Try to narrow it down to just one. The point is to be concise but 'triggering', so that at the end of this process you have formulated one sentence like this (that we refer to as the 'action sentence'): "David buys ready-made food when he is alone."

Now take a very large sheet of paper and write the sentence down right in the centre. It would be even better to use pictures or drawings describing the sentence, but please be careful to use clear images that everybody in your team can understand, and try to avoid clutter.

"Simplify, Simplify, Simplify. It's not an art to speak to people from a platform from above — talk to people where they are. This is how you connect to the 'real world'." — ANNE LISE KJAER

PAST ACTIONS  CURRENT ACTIONS  FUTURE ACTIONS

IMAGINE HOW RELATED ACTIONS HAPPEN IN TIME

This step is the start of the process of framing your reflections from the previous steps. Now that you have a central point, think of the most important events that happened before the main action (past), that happen during the main action (current) and that will happen after the main action (future). Write down these events on a timeline, starting from the centre and moving to the sides. To keep it all on the same abstraction level, try to use only verbs to describe these actions, like 'getting a phone call' or 'sleeping in the car'. This is particularly useful when there's too much unstructured information as it helps you to focus on the 'actions'.

It might be that while you were exploring the context, a tight schedule didn't allow you to observe and reflect on important events that happened previously or afterwards. Don't worry, just use your common sense and your empathy skills that you've been exercising up till now. A tip to help you spot the really relevant events is to ask yourself not only about what happened but also about what the main actor's aspirations were and what he or she did, does and will do in order to achieve those aspirations. Actually, one of the most difficult things designers attempt to do is to predict people's behaviour. Sometimes you get it right, sometimes you don't. We could try to explain human behaviour as being closely related to our aspirations: we do something because we would like the future to be somewhat like our expectations of it. Aspirations are our 'future experiences' that arise from what we feel and what we

VISUAL THINKING

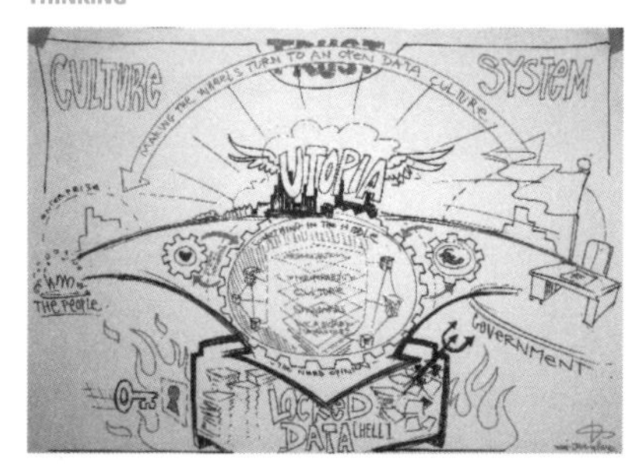

do at a certain moment and what we expect from the future. Therefore, it might help you to attach an 'aspiration note' next to each event to remind you of the way the main actor might be feeling and what the goals are during every event.

Recommended techniques: since you will use this step to start framing everything you have observed up till now, you might need some tools to help you do it in a way that is inspiring and triggering for the next steps too. So, first decide what kind of technique will help you visualize the network. You could use breakdown techniques, system mapping techniques or mind mapping techniques. Mind mapping is a visual and verbal technique usually used to structure complex situations in a radial and expanding way. A mind map is a pattern of related ideas, thoughts, objects, people, etc. There are also mapping techniques that help you highlight direction, importance, cause and effect, etc. These can be in the form of a tree or a flow chart. Actually, we recommend giving it a lot of thought when choosing a technique before actually plotting the network. What works best for us is to mix them up so that we can create a journey map where we start from a central point, depicting relationships and a time factor.

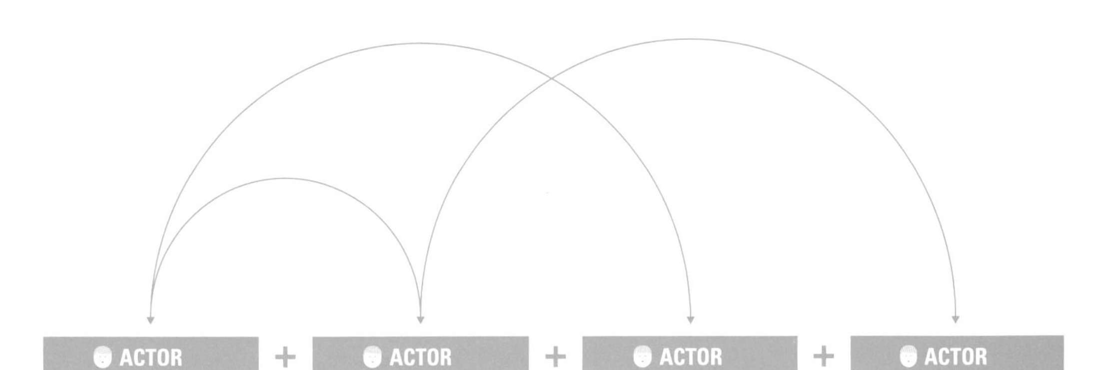

IDENTIFY ALL OTHER 'ACTORS' AND SHOW THEIR RELATIONSHIPS

Now it's starting to look like a network! Think of all the 'actors' involved in the actions you've just placed in the timeline. The actors could be people, spaces, organizations, products or services — as long as they do something (perform an action) and the resources of value may be virtually anything they willingly exchange. It may be knowledge, other products, other services, emotions or money. Connect actors and actions with lines and draw arrows to show how they exchange resources of value. Label the arrows so that you understand the flow of exchange. Try to highlight with different colours so that you can cluster important relationships.

So now you've become familiar with the context. You've pictured the network as it is at the moment and perhaps you're wondering why it should change, what kind of change that might be, and if the change actually means an improvement. If these are the kinds of thoughts running through your head then you're on the right track! The process you're probably going through is called 'abductive thinking', which takes place once you get to see the whole. Abduction has been described as 'the logic of what might be', 'the argument of the best explanation', but most of all, it allows the creation of new knowledge and insight to explain the relationships. This is when you begin 'reframing' the network by revealing new relationships between elements that are apparently unrelated. Answering the question of why you would want to generate change in the current network requires a great understanding of the motivation of the actors who want to exchange value. To help you understand this motivation there are several tools you could use and we will present them in the steps below.

"...complexity and the excess of meaningless choice is still an unsolved issue in most developed western economies. Therefore intelligent reduction will play a key role in the future of innovation and design. Creating successful seamlessly integrated products, services and experiences in the future will mean changing the way we think about people. Concepts simply must become people-centric, demonstrating empathy for the cultures they serve and respect for the context in which they exist. A challenge within this is achieving consistency throughout both intangible and tangible elements." — ANNE LISE KJAER

Network Focused Design

- **A** Explore the context in actions
- **B** Imagine how related actions happen in time
- **C** Identify all other 'actors and show their relationships
- **D** Identify challenging relationships
- **E** Reframe the network
- **F** Define your criteria & visualize the new network
- **G** Identify the communication flow
- **H** Design the touchpoints
- **I** Test

PHASE 2 SETTING A DIRECTION

Here, you will question how the information flows work and how people exchange their resources. When you start to reframe the network, you will be actually setting the design direction of all its elements, you will be creating your own design guidelines or the design criteria for the whole network.

In the following three steps you will be guided to confront your ideas to come up with new ones and set a new design direction.

IDENTIFY CHALLENGING RELATIONSHIPS

Take another good look at the whole network. It's possible that you may have already detected some 'problems'. But you may also spot some opportunities to reframe the network without there actually being a 'problem'; that's what we call 'challenging'. This could mean adding new elements to the network, or changing the way the actors exchange value. Think of the cause-and-effect relationship between aspirations, actions and the value exchange. It's possible that there will be many challenging relationships, all scattered throughout the network you've visualized. If that's the case, then try to highlight them so that it's obvious which directions the arrows are coming from and how the elements are related. Then synthesize the challenging relationships into a 'How to' question (H2question), remembering to keep the focus on the main actor. For example: How to provide informal healthcare to my grandmother and strengthen our personal relationship? Don't worry about the grammatical structure, H2questions always sound a bit weird, but they tend to simplify situations. Besides, once you've created the H2question you're halfway reframing the network. A H2question is always open to many answers and it focuses your mind on the future possibilities.

"If you only have a hammer you will see all your problems as nails." — ABRAHAM MASLOW

Recommended techniques: to identify challenging relationships, there are some techniques you could use. Sometimes the challenging relationships 'pop out' from the network by using just your common sense, but sometimes they're not that obvious. In those cases, try using wishful thinking, absence thinking or worst case scenario and/or forced conflict techniques. There are plenty of books and online sources you could refer to. We recommend the book Creative Facilitation by Marc Tassoul to help you with this and with the next step. Game Storming by Gray, Brown and Macanufo or the Facilitation Guide Series by David Sibbet can help you generate and organize ideas in a team.

HOW TO IMPROVE ACTIONS BETWEEN ACTOR  ACTOR

REFRAME THE NETWORK

We chose the term 'reframing' because it makes us think of looking at things from a different perspective. If you simply check the term on Wikipedia, in the context of management and marketing, reframing means giving a new value to a product/service by finding a new market/context. This sounds fine, but actually what we mean by reframing is finding new relationships of value. This means that there might be new actors, new resources, or simply new ways of exchange.

In the previous step you already started reframing the network in some way. Even though you may think you already have the answer, you're only halfway. Now's the time to start generating lots of ideas!

This is where your H2question comes into play. You can use some creative techniques to get as many ideas as possible to answer the H2question and enable you to reframe the network.

Recommended techniques: this step should be a boost to your creativity and should help you think 'out-of-the-box' and to challenge the ideas you already created in the first steps. You might go back to those ideas after this step but that's OK as long as you truly tried to confront/evaluate them before actually choosing those initial ideas. So now you'll have a network filled with challenging relationships and a H2 question. These are useful in helping you generate new ideas. But there are also idea generation techniques to help you a bit more. Choosing the right techniques depends on the time available and what you and your team might feel more comfortable with. For example, some teams prefer intuitive techniques such as 'guided fantasy' or 'future perfect', in which you imagine a fantastic scenario that answers the H2 question while the rest of your team frequently says 'why?' to your narration and you have to provide answers. Then another member takes the narrator role and the team members continue asking 'why?'. All the answers are written down in short sentences by an assigned note taker so that you can categorize them and choose the best one for the next step. There isn't really a right or wrong in this step but you have to have clear in your mind that whatever technique you use is intended to confront your existing ideas. You could also try brainstorming or brain drawing to answer the H2 question by building ideas on top of the H2 question. For example, the H2 question: "How to restore the traditional feeling of dinner-time to David?" Try to divide the H2 question into the core parts: Restore — traditional feeling — dinner-time. From here on you can use these core parts to brainstorm or brain draw. Of all the techniques available, we recommend force-fit techniques. To force-fit, draw a circle on a large sheet of paper and divide it into the same number of parts (pie slices) as there are team members. Then every member writes down on a Post-it one idea about 'Restore' and sticks it onto his/her part of the paper. Let's say one idea written down on a Post-it is 'remember'. Then the sheet of paper will be rotated clockwise so that each team member will see the idea written down by the member to his/her right. For example the team member next to you writes 'remember' and you build upon his/her idea with something like 'remember + smelling' and then the paper is rotated once again, and so on. The point is to use each core part of the H2 question to generate ideas.

CREATIVE SESSION

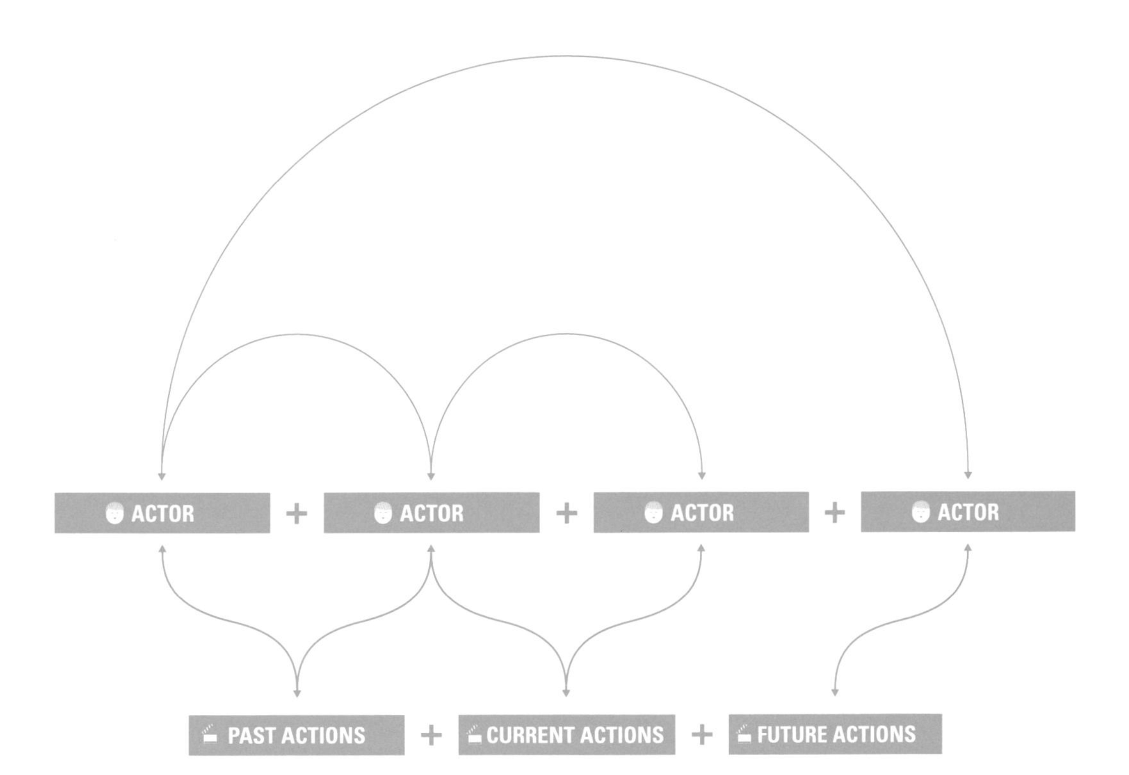

DEFINE YOUR CRITERIA & VISUALIZE THE NEW NETWORK

Once you've collected all your ideas, it's time to select the best one so you can visualize it in the new network. When you go on to this step, don't forget to keep the main actor at the centre of focus. And be sure to have the material you've collected (personas, current network, the how to question, etc.) at hand and always within view.

"You should be aware of where to place the boundaries and how they are connected. How many layers are there, where is the control? How many people? Is it only machine to machine? Without any human beings involved?" — JAN BUIJS

A handy way to define your criteria for selecting the best idea is to identify whether the actors are motivated to exchange resources. Start clustering them, for example into 'very important actors', 'important actors' and 'not so important actors'. Then think about why they would be motivated to be part of the network and exchange resources. It could be that the way to exchange resources is the most important element of the network or in other words, the way to exchange resources is what makes the network possible. Think of speed, real-time, intuitive properties as examples of ways to exchange resources. Try to identify whether the relationships you are proposing help them meet their aspirations.

"There are always limitations, let's say money, time or your own knowledge or the team you're working with, your network; all these limitations together build up your framework. You should go beyond what you thought you could do within this frame."

— HARALD DUNNINK

A useful tool to help you define your criteria would be to identify where the costs are and where the revenue flows are in the relationships. This could help you predict whether there's a potential business case in the idea you ultimately choose.

Prepare all the materials (actions, actors, resources, ways of exchanging resources, costs, revenue flows) from the best idea selected to visualize the new network. The new network will show new relationships and the value exchange between new and old actors. Use pictures, colour codes and aspiration notes when necessary.

"The classic usability definition is basically operational, like how long does it take, how many hours do people make, you know, very strict. 'Ease of use' goes one step further, it means that when people look at something or encounter it for the first time, they can just walk at it and start using it. But now we are more into the design of total ecologies, in which there are totally different services and different devices. You might not even use all of them, maybe only one sometimes. And that is why 'us-ability' became rather mundane as a term, and we needed a new term, so now we got 'experience'. If you look at the professional domain of user-system interaction and you want to make an 'experience' operational and you want to evaluate it or test it, then you come back to the old usability type of things. Which means we just broaden up our design approaches. For designing, I think designing for 'experience' gives you more freedom to be more experimental and creative. I mean, when you design only for usability, your constraints are strict, and sometimes they have to be strict. If you talk about medical systems or an airplane, you have matters of life and death, so you have to be strict. But at the fuzzy front end of innovation or if you design something that people using it will judge whether it is successful or not, de-signing for 'experience' would be more appropriate. To create an 'ideal experience' when it is individually different is quite hard. I think you have to give people options to create their own experience within the environment you 'provide'."

— MADDY JANSE

The next step is where everything comes together. Here you will have to translate the new network into a Meta Product. This means finding the match between the information streams, proper interfaces, interactions, materials and technology.

Also, this is the crucial point of an iterative process. Testing, revising your guidelines and criteria, and testing again. The previous steps helped you to gain empathy and to be sensitive to the actors' aspirations so that you could reframe the network with new relationships of value. By now, you have grasped how resources can be exchanged in a better way. You already have a rough idea of new devices, new actors, new services and business models and new streams of information that will bring the network to life. However, so far you've gone through mostly holistic tasks. Now it's time to zoom in and design the interactions.

CO-CREATION SESSION

Network Focused Design

- **A** Explore the context in actions
- **B** Imagine how related actions happen in time
- **C** Identify all other 'actors and show their relationships
- **D** Identify challenging relationships
- **E** Reframe the network
- **F** Define your criteria & visualize the new network
- **G** Identify the communication flow
- **H** Design the touchpoints
- **I** Test

PHASE 3 DESIGNING THE META PRODUCT

So far you've visualized a new network but not yet all the interactions. This stage is the beginning of detailing the Meta Product, and you will need to collaborate with other professionals intensively and work in an iterative way to make sure your guidelines are correct.

"...actually, a designer does not design the service per se. In my opinion the designer sets the network up and the people that act in it design their service activity."

— FRIDO SMULDERS

After exploring and generating, now it's time to build upon the best idea, which by now is the new network. Now we have to zoom in to define each element and every relationship, and zoom out to check the coherence of the whole network.

The following three steps are a general guide to help you take off for designing the interactions of the Meta Product.

 +  + 

IDENTIFY THE COMMUNICATION FLOW

Take a look at the new network you just created and answer the following questions:

1. What could be 'raw data'? This is the information that could be sensed, tracked, measured, controlled, categorized and interpreted. How is this data retrieved? Is it by conscious decisions of the actors? Or should it be totally automated without any human intervention?
2. What kind of 'interpreted data' would there be? This is the interpreted information that is accessed by certain actors for a particular purpose.
3. What kind of touchpoints are necessary? These are the media where the raw or interpreted information is retrieved or accessed. Depending on the new network (your design guidelines), these touchpoints can be devices, web interfaces, spaces or existing products.

"Technology is just a means. It is actually what we do with it that counts. The distinction between online and offline is not what counts. In some situations people want to feel that they are in charge and take conscious choices. So if they know what the device can do for them and what they can do with it, whether it is online or offline is irrelevant for them. But indeed some functions have to be transparent nowadays precisely because people want to feel they are in charge, and so you have to give people the choice of interacting with the device or function or not."

— ANNA VALTONEN UMEA

"Embedded electronics in materials is just a reality, whether to use connectivity in objects or not is just a choice or a design direction to decide. But they should be made very carefully and critically by understanding the human dimensions of their actions... you should design technologies for humans and not humans for technologies... it sounds like a cliché but is still very true." — FEDERICO CASALEGNO

TOUCHPOINT 1

TOUCHPOINT 2

TOUCHPOINT 3

TOUCHPOINT 4

DESIGN THE TOUCHPOINTS

As explained previously, the touchpoints are the media through which raw data and interpreted data are retrieved or accessed in the network. By now you have decided the type of touchpoints you will be implementing in the network. Now you have to design them. These touchpoints will be used by the actors you've already defined, so follow your guidelines and the criteria you set in G). In other words, this step is all about designing interactions in detail.

"Take human touch for example. This is one of the most natural interactions of human beings. We try to mimic it in many ways because touching is an interaction that is so deeply embedded within us that if we can't have it, we will mimic it. But there are other types of interaction that people are willing to learn or to develop, new ways that are not always what you would expect. We could see this with the arrival of haptic technology, where you can actually feel a key where there isn't a key only because for some people it just feels more natural to press something physical although the physical doesn't actually exist. The Internet of Things can be similar to these situations: a physical interaction might be more natural to the user in some cases, although the physical thing might not be needed for the interaction. However I would say that rather than focusing on what is physical or non-physical, I would focus on what is the most natural way to interact for people, and then start building it from there." — ANNA VALTONEN

Here you have to make careful decisions for each interaction:

1. What kind of perception is the best? Is it visual? Or is it aural? Presence, smelling, squeezing, pressing? Or are there imperceptible interactions?
2. How high a level of learning or engagement does the interaction demand from the actors?
3. What kind of emotions are intended to be experienced?
4. Are the aspirations of the actors in line with the goals of the interactions?

"...the people designing at a 'network level' will have to be clear about when, how and why the physical or material parts of the service have to connect to the digital parts and where the boundaries between different partner organizations can or should be located." — LUCY KIMBELL

Recommended techniques: it's a challenge for designers nowadays to communicate all the layers of interaction a Meta Product has in the early stages of the design process. Factors such as time, number of interfaces, social serendipity, multi-users and so on are difficult to plot on 2D drawings. The challenge is to communicate concepts in a clear way and yet quickly and at low cost. The first step to designing the Meta Product would be to have a clear visualization of the new network and the communication flow diagram that everybody in your team can see and understand. Then make use of quick & dirty tools such as video-taping role-playing and mix them up with storyboards or infographics. Prepare the role-playing with cardboard/paper movable interfaces, allow enough time for the various activities, and try to be precise about the dimensions of devices and spaces. Try to add details when necessary. For example if the main idea of the concept is about 'pouring water' then try to do it with water. Or if colours are important then apply colours. You can decide for yourself the level of detail that's useful for testing in the next step.

STORY BOARDS

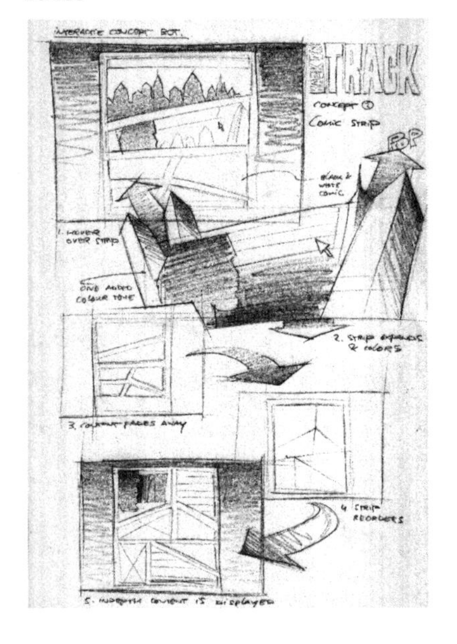

MAIN ACTOR + MAIN ACTION

TEST

This is one of the most dynamic parts of the design process, where you will see if you got it right. It's also a crucial step before making investment decisions. First of all you have to define clearly the elements that are testable, the ones that can be left out of the testing process, and the ones that can't be tested at the moment. Once you know exactly what you want to test, you can choose the best way to prototype your Meta Product. As mentioned before, communicating or testing concepts at this point of the design process does not have to be costly. In this context, check out the work of Berg Studio from London[9]. In 2009, Berg undertook a project of shifting paper magazines to digital media. They did the prototyping and testing in a very fast way with low-fidelity sketching. They taped the pieces of magazine together in a line, hung them from a door frame and then filmed it with an iPhone. They could reflect on the results and pass them around to get a rough feel of what it might feel like on a hand-held device. This is a very quick and dirty example, and it depends on the time and resources available which testing procedure is feasible. The key element to testing effectively is to be able to make associations from the results and look back to the previous steps to refine the new network and the interactions.

MAGPLUS

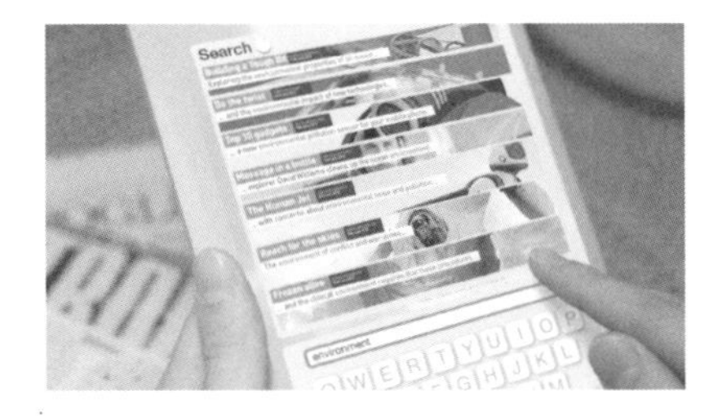

NETWORK
FOCUSED
DESIGN SESSION

CHAPTER 5 — SUMMARY INSIGHTS

1. Describing, interpreting and imagining the future are the key activities of a reflective practitioner. So they are also key for anyone who designs.

2. The reflective practitioner is someone who learns by reflecting on current experience and applies that learning to future practice. Holistic most of the time, analytical at other times; intuitive most of the time and systematic at other times, all in order to find and frame the problem together with its solution.

3. The reflective practice is an unstructured self-regulated approach that aims to help understanding and learning. It is very similar to any design process. Except that designers rely heavily on their intuition to innovate. Designers use 'abduction reasoning', which is basically the 'logic of what might be'.

4. The golden rule 'knowing your materials' is a reflective process to discover what can be designed, why it can be designed and how it can be designed.

5. Information and connectedness are the new materials of designers, as these are 'the fuel' of the interactions we design.

6. Designing a Meta Product can involve considering multiple users, multiple services, and multiple products at intermittent times, some elements are visible while others are invisible, some elements are human-initiated while others are automatically generated. Hence, designing Meta Products requires a network mindset, transdisciplinarity and great communication skills from team members involved in a project.

7. Designing systems is slightly different from designing networks. The difference lies in their nature and in what they trigger in you. A system tends to be closed and a network tends to be open. A system might require a more analytical way of thinking and a network a more holistic reflecting process.

8. NFD is an approach that will help you to identify opportunities where new relationships of value can be created. This approach focuses greatly on taking you 'out' of the technology world for a while before actually designing information flows and interactions. It also guides you to think in networks and to focus on the relationships that build them.

FITBIT

Fitbit is a Meta Product that measures your daily activities. It keeps track of calories burned, number of steps taken, distance travelled and sleep quality. By setting up goals and reflecting upon your actual results, you can improve on your health.

The total Fitbit network consists of three parts: a clip-on device, a base station and a web-based software application. The clip-on device has a 3D motion sensor (like those in the Nintendo Wii controller) and stores motion data. Whenever this device is near the base station, it silently uploads the stored data. Besides automatic tracking, you can also log activities yourself. The combination of automatic and manual input gives a complete estimate of
your activities.

In the online environment, a training programme reviews your current activity level and creates a personalized fitness plan that recommends you to gradually increase your movement. You can follow this plan and see how you rank against your peers along the way. It works by setting up a custom tracker to easily identify and monitor areas you'd like to improve. Maybe you like to monitor your daily commuting time, your caffeine intake, or your sleep patterns. You can set goals for your own healthy lifestyle and follow the recommendations that the Fitbit training programme gives you.

According to Fitbit, tracking is the first step to changing your lifestyle. This notion is precisely the value of its web platform. Without the platform, you would not be able to have a concise visual overview of the long-term trends of your behaviour.

Fitbit is intended to be worn day and night. The product itself is small, but not tiny. If you want to monitor your sleeping patterns, you need to wear it, for example, clipped on a bracelet.

Fitbit is a great example of quantifying yourself. However, we believe it raises many questions concerning the meaning people will assign to these new interactions, and the possible implications or effects of self-monitoring. There are many self-monitoring products on the market today. Sure, they claim to help you achieve a healthier lifestyle, but could the fact that you're continuously conscious of your behaviour also be a source of stress? There are many important factors to consider when designing these types of networks in order to trace the impact and relevancy of new interactions on people's lives.

IMAGES PROVIDED COURTESY OF FITBIT, INC

NOKIA X BURTON
PUSH SNOWBOARDING

Push Snowboarding is an open snowboarding innovations platform and is the result of a joint effort between Nokia and Burton. They have created a hardware-software kit aiming to provide insights into what happens both physically and mentally to snowboarders when they are performing. The current kit consists of sensors collecting data from five key measurements (heart rate, pressure, speed, orientation and rush) of six different pro snowboarders. The sensors are hacked together from parts that interface with an Arduino board. The Arduino board basically acts as a wireless bridge between a Nokia phone and the measuring devices. A Nokia phone then finally stores all the collected data in a centralized place.

What makes this Meta Product interesting is that it is open; as a developer, you can download the latest source code and hardware diagrams to build your own version of the kit. You can also submit ideas and improvements to the project from your own experiences. The choice for an Arduino board is appropriate, because it is specifically designed to be a simple and easy-to-use hardware board with which you can hack together your own devices. Nokia & Burton state that they are still refining the technology, but they believe that innovation comes from openness, and that by co-creating with snowboarders, they can enhance snowboarding in ways that matter most to riders.

This example of co-creation shows that products can find their meaning and value by involving the right stakeholders during the process of development. Nobody knows if the Push Snowboarding project will end up in a successful commercial project. It could just as well fail, or never even be taken further than the current state of a hacked-together project. But the process of finding new meanings and discovering value by developing out in the open is certainly interesting. Do you have any ideas about improving this project? How about turning this into a snowboard learning tool? Or a professional training tool for the Olympics? Or a snowboard game between friends? What do you think?

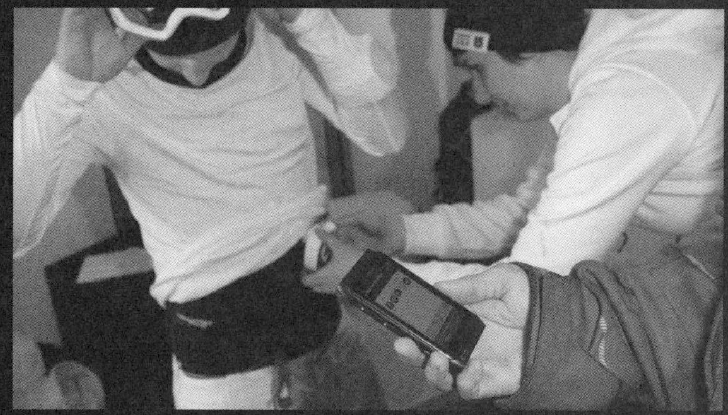

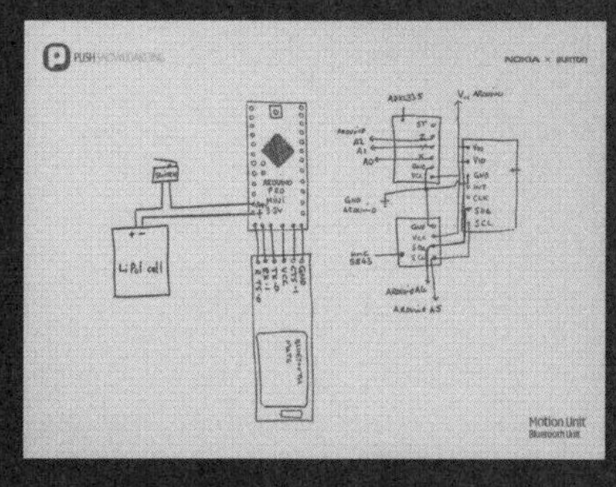

IMAGES PROVIDED COURTESY OF NOKIA, INC.

How would you define an interaction?

JENNY DE BOER
TNO

I think an interaction is about what you do with something else. In co-design we let people interact with certain technologies so they can experience what the technology can do for them. For example, we gave a demonstration in Namibia of a basic interaction to a group of people and then we let them do it themselves. They could make a text message and then see it on the web for example, and once they have that experience they are able to think how they can apply it for themselves.

Do you have some important considerations or principles when designing interactions involving the web or other ubiquitous technologies?

JEROEN VAN GEEL
Fabrique

Context: it really depends on the environment in which the product or service is used. You don't always know this, and even when you do, it can be difficult. So good research is really important here. When you design for an airport you need to observe how people behave. When you design a mobile device you have to test all possible scenarios. Then there's also content: you can design the best digital product ever, but if the content is bad it will not work. Content is more important then anything else. Ugly websites with great content are being used, great looking websites with crappy content are being ignored. So from the start you have to work together with the client and understand what content they have or help them create the good content. Lastly comes design: what we design and how we design it. If it is something totally unexpected, new, or a pattern people recognize. This is something you have to do right. And knowing your users helps. Telling the right story in the correct language.

MADDY JANSE
Philips Research

I have a very abstract definition; an interaction is between two entities, two structures that have a relationship with each other that dynamically changes. So an action by one of the elements causes a reaction by one of the others, that causes a reaction with others in return and it goes on and on like this. Whether it is a technical interaction between two software systems, or whether it is an interaction like communication in human exchange or whether it is between humans and systems. It's always about these reactions of two behaviors which are always dynamic in time, it's never static.

FRIDO SMULDERS
Delft University of Technology

Since interactions are everywhere and at any time, it is good to differentiate between intentional and non-intentional interactions. Intentional interactions have a specific aim and are maybe better called interventions, because they have the aim to bring about a certain change, be it in feeling, information or experience. Within the field of human interactions one can't speak anymore of sender and receiver. Sender and receiver continuously change roles, and even these roles are operational at the same time, like body language. Products and systems with intelligence aim to have some sort of pre-understanding of what the human user would like to receive or experience. This is not far off from what we call empathy, so empathic products are able to 'read' our feelings and maybe hidden needs and react on that. For Meta Products this could be specific parts of the total network that provides this empathic intentionality in a seemingly blurred system of interactions where sending and receiving happen at the same time. Question is, how to design all this?

ANNA VALTONEN
Umeå University

We have to think of interactions that just feel natural to people, intuitive. What is the most human interaction? There are some things that just feel more natural to interact with than others. If we eventually use only 'black boxes', like the smartphone, then designers will 'fake' the human interaction in some way. They can hide the black box and create a 3D avatar in front of it that you can actually hug as if it was a physical person. There are many possibilities, we are getting there. The meaning and humanness of that meaning of the interaction is what counts, whether it is a physical box or a piece of software.

FEDERICO CASALEGNO
Massachusetts Institute of Technology

The way you design something from the global village perspective is very different - which is when you design something which assumes the interaction happening remotely - versus when you design something for the fractal village - which assumes the interaction is happening in a very close network of actions. When you design 'simple' interactions (without embedded electronics) and the focus is on the fractal village, then physical materials have an important role. The fact is that five years ago you wouldn't have designed an object knowing that it can be so immediately connected, recognizable, and that the matter of this object can change and react. Today, this is possible to have. So the way of designing is different, you have to consider a new spectrum of variables. Also, a very critical mind towards too many variables is crucial.

JAN BUIJS
Delft University of Technology

Designers should look at the total system and then see what kind of communication or information streams are necessary or interesting between the system elements. And if there is an opportunity for the Internet of Things to make those streams better, then consider what that would actually mean. Designers should be aware of where to put the boundaries in the system and how elements should be connected. How many layers are there? Where is the control? How many people? Is it only machine to machine? Without human beings involved? People assign meaning to these technologies, which is translated in their behavior, and this is exactly what a designer has to look at.

FINAL THOUGHTS

If we have achieved our goal with this book, you should now be wondering what type of Meta Products could be meaningful to your client, to your processes and to your everyday life. You will be asking yourself what type of networks you will need to design, how you are going to motivate your stakeholders, and what you have to do to make it happen.

We all know that day-to-day business life runs at a lightning pace. Which makes it all the more important for designers in particular to take time out to reflect. In the end, it's designers who have a direct influence on everything out there in the world. We hope the insights in this book will help you further in your profession. We trust too that you have been inspired to identify possible innovations by looking at how people assign meanings, build their aspirations, learn, construct knowledge and adopt change in our connected world.

This is the first time we've written such an extensive overview of the current situation, including both human and technological perspectives of what has led to the present moment and what might lead our products and services in the near future. We admit it was ambitious. We blended together sociological perspectives, historical facts, cognitive science and psychology literature, business and marketing principles, and design theories into one, with the aim of providing a holistic view of the phenomenon of Meta Products. We believe designers need to become network-focused without losing touch with their expertise. Moreover, forward-looking organizations, firms and institutions are increasingly realizing that they need expert designers who have a great deal of empathy and strategic skills to deal effectively with the complex nature of the web-enabled product-service networks that will be filling the environment in the near future.

This book is an on-going project, the beginning of a dialogue. There are still so many questions to be answered. Just thinking about the impact the new interactions will have on our lives and communities will evoke a lot of questions in many different fields. For example: what will happen if people adopt 3D printing on a massive scale to make their own products? Or if the clothing we are wearing can communicate with our doctors? Or if we stop owning houses and replicate our own environment digitally wherever we go instead? Attempting to answer these questions is both exciting and challenging. We cannot be entirely sure of how to answer them at this moment, but we are sure that designers have an important role to play in helping society to do this; not only in an effective way but also with empathy for and responsibility towards our fellow-inhabitants on this planet.

We would like to continue the dialogue on Meta Products on our blog www.metaproducts.nl. This is where you can share your reflections, examples, and stories, as well as download materials and get updates on Meta Product topics. We are really curious and excited about the contribution Meta Products will make to society as a whole. Let's connect and share ideas!

ANNE LISE KJAER
Kjaer Global

Anne Lise, CEO of Kjaer Global, breathes life into trends forecasting and champions 21st-century approaches to sustainable strategy and meaningful innovation. Her pioneering multidimensional approach combines research-based methodologies with informed intuitive understanding of tomorrow's people. With a long track record of assisting leading brands successfully navigate the future, Anne Lise is an in-demand speaker at conferences and events worldwide. She has also delivered lectures for Central Saint Martins, Copenhagen Business School, ESADE, Lund University and Royal College of Art.

LUCY KIMBELL
University of Oxford

Lucy teaches an MBA elective on design at Said Business School, University of Oxford since 2005. She has her own consultancy Fieldstudio, and she is associate at TaylorHaig (London) and The Policy Lab (Boston), both working to innovate in public service design. Recent keynotes include the Service Design Network (2010), Design Management Institute (2010), and Ethnographic Praxis in Industry Conference (2008). She publishes in peer-reviewed journals as well as trying to find ways to make academic work more digestible to wider publics.

ANNA VALTONEN
Umeå University

Anna Valtonen is born in Helsinki, Finland. Since 2009 she is the Rector of Umeå Institute of Design (UID), Umeå University, Sweden, one of the top industrial design educations in the world. Before this position she worked 12 years for Nokia, lately as Head of Design Research & Foresight. Anna Valtonen's research interests are in design, its history, and how companies and nations can best use design for their competitiveness. Her PhD, from the University of Art and Design in Helsinki, Finland (currently Aalto University) was on the professional development of industrial design practice. She has also been active as a researcher at the Department of Strategic and Industrial Design at the same university and as visiting professor in the Department of Management at ESSEC Business School Paris, France. Valtonen also holds many positions of trust within universities, organisations and corporations.

FRIDO SMULDERS
Delft University of Technology

Frido holds a PhD in Innovation Sciences and a Master degree in Aerospace Engineering, both from Delft University of Technology. Presently he is Associate Professor Management of Product Innovation and Entrepreneurship and the Director of the MSc programme in Strategic Product Design at the Faculty of Industrial Design Engineering (IDE), TU Delft. Prior to joining IDE Frido was working as a management consultant in the field of innovation, creativity and technology. He lectures on Project Leadership, Leading Strategic Design, Creativity, Entrepreneurship and Collaborative Business Design. He has published two books and over 60 articles on topics including product innovation, creativity, fuzzy front end and collaborative innovation. His research focuses on understanding the social (people) dimension of innovation and design. With this research he aims to connect the engineering sciences to the social sciences.

JAN BUIJS
Delft University of Technology

Jan has been full professor and chair in product innovation and creativity at the Delft University of Technology since 1986. He got his MSc in product design and his PhD in management sciences. Before working at the Delft University he spent ten years as a management consultant. As a professor he teaches and studies product innovation, creativity and multi-disciplinary innovation teams. He was president of the European Association for Creativity and Innovation (EACI) for many years.

FEDERICO CASALEGNO
Massachusetts Institute of Technology

Federico is the Director of the MIT Mobile Experience Lab and Associate Director of the MIT Design Laboratory at the Massachusetts Institute of Technology. Since 2008, he has been director of the Green Home Alliance between the Massachusetts Institute of Technology and the Fondazione Bruno Kessler in Italy. He is adjunct full professor at IMT Institute for Advanced Studies Lucca, Italy. A social scientist with an interest in the impact of networked digital technologies in human behaviour and society, Dr. Casalegno both teaches and leads advanced research at MIT, and design interactive media to foster connections between people, information and physical places using cutting-edge information technology.

MADDY JANSE
Philips Research

Maddy was Principal Scientist at Philips Research from 1987-2008 and currently adviser. Worked as specialist in the domain of user-system interaction in different research groups at Philips Research. Project leader and coordinator for several EU- ICT FP projects, i.e., AMIGO — ambient intelligence in the networked home environment, NexTV and ICECREAM projects in the interactive media domain, editor vision document for the NEM project (networked electronic media). Managing Director of two-year post-Master's program in user system interaction at the Technical University Eindhoven in the department of Industrial Design from 1998-2011. Ir (M.Sc.), 1967, Wageningen University, Food Science and TechnologyPhD, 1983, University of Minnesota, Minneapolis, US, Cognitive Developmental Psychology, Problem Solving Behaviour.

ALEXANDRA DESCHAMPS-SONSINO
Designswarm

Alexandra is a product designer & entrepreneur. Born in Montreal, she grew up in Paris and the Middle East. She completed a B.A. in Industrial Design at the Université de Montréal and a Master's degree at the Interaction Design Institute Ivrea (IDII). She worked on digital & innovation strategy for clients like Blast Radius and startups like Blyk, Jaiku & ThingLink. She co-founded Tinker, a smart product design studio, which she ran from 2007 to 2010. Focused on creating product experiences that linked the digital to the physical, some of Tinker's clients included National Geographic, BT, BBC, The Evening Standard, Whirlpool & Wired. She is now a consultant on 'internet of things' issues in product & digital design, an evangelist for Lirec (an EU-funded project about robots) and partner in design partnership RIG London.

JEROEN VAN GEEL
Fabrique

Jeroen is senior interaction designer at Fabrique [brands, design & interaction], a multi-disciplinary design agency in the Netherlands. He has many years of experience in the field and works for major clients such as D-reizen, Rijksmuseum, 9292 and Schiphol Airport. Jeroen is founder and editor-in-chief at Johnny Holland, the online magazine around experience design.

JENNY DE BOER
TNO

Jenny is educated as an industrial designer at TU Delft with a Master's degree in Design for Interaction. Since April 2007 she has been working at TNO on research and facilitation of co-creation in ICT innovation processes. Since 2008 she has a special focus on co-creation processes for the successful development of mobile phone services in developing regions. She has initiated and participated in various projects all over Africa for industrial, governmental as well as non-governmental organizations.

HARALD DUNNINK
Momkai

Harald is a relentless lover of graphics and design. After finishing his graphic design studies in 2002, Harald (1981) decided to team up with his then partner to found the digital creative agency Momkai. While at first being more focused on visual communication he invited his current business partner and technical director Sebastian Kersten to join the agency and hence give it a proper technical foundation. Gradually they moulded Momkai into a renowned studio for high-end interactive projects. Harald has final responsibility for all projects as creative director. He combines this function with that of designer, mostly focusing on graphical user interfaces and from time to time illustrational work.

SEBASTIAN KERSTEN
Momkai

During his studies Sebastian Kersten (1976) developed an interest in user friendly dynamic systems that allow users to create and maintain interactive projects with no or little technical knowledge. This resulted in 2001 in the development of the first versions of MimotoTM CMS which is currently growing into a framework that is able to create rich internet applications. After joining Momkai in 2003 as Technical Director he and his business partner Harald Dunnink build Momkai on the combination of design and technique to create high-end interactive projects like Nalden.net and Pete Philly.

MARC FONTEIJN
31Volts

After graduating at the University of Applied Science in Utrecht, Marc started a mobile development company back in 2004. After two years in progress, it became clear that the real interesting challenge isn't technology. It's figuring out how to deliver services that enrich people's life. So in 2007 Marc partnered up with Marcel Zwiers to found 31Volts, the first Dutch service design studio. Marc also sparked mobile innovation as the chairman of Mobile Monday Amsterdam. Curious by nature and fascinated by things that say beep. Marc has connected a fish to the web at beslisvis.nl

TIM VAN DEN DOOL
Innoviting

Since the first time Tim heard about technologies like RFID and GPS he was interested in connecting the physical space to the virtual world. Tim van den Dool set up his company INNOVITING around realizing solutions with these techniques. He developed projects from introducing self-service in libraries up to behaviour monitoring of the elderly. All the projects were based upon the W5 principle. This means being able to always answer the following questions: when, where, who, what and most importantly, why?

Actor
Active element of a network that has the ability to exchange resources. These resources can be money, time, concepts or anything that is valuable for exchange purposes.

Actuator
A mechanical device dedicated to move or control a mechanism. It can be operated by a source of energy, usually an electric current converting it into some sort of motion.

Ambient Intelligence
Dedicated environments formed by devices and other elements in the surroundings that are embedded with technology and aim to function intelligently and dynamically towards the presence and actions of people.

Application Programming Interface (API)
A set of rules and specifications that software programs can follow to communicate with each other. It functions as the interface between software programs and facilitates their interaction.

Arduino
An open-source single-board microcontroller dedicated for quick-prototyping purposes of the interaction between code, sensors and actuators. It is designed to make the process of using electronics in multidisciplinary projects more accessible.

Artificial Intelligence (AI)
The study and design of intelligent machines and systems that perceive the environment and take actions accordingly.

Aspiration
Internal personal drive or force to accomplish something.

Augmented Reality (AR)
A term that refers to the direct or indirect interaction with physical elements or real-world environments that are enriched with computer-generated sensory input such as sound, video, graphics, GPS data, or by any other extra digital layer of information.

Cloud computing
A model for enabling ubiquitous, convenient, on-demand network access to a shared pool of configurable computing resources (e.g., networks, servers, storage, applications, and services) that can be rapidly provisioned and released with minimal management effort or service provider interaction. (National Institute of Standards and Technology)

Co-creation
The level of involvement of other relevant stakeholders in the design process. It can include a series of techniques for design practices in which the stakeholders participate actively or passively in the design process and guide some decisions.

Cognitive load
Used in cognitive psychology to describe the load related to the operational control of working memory.

Context mapping
A procedure for conducting contextual research with users, where tacit knowledge is gained about the context of product usage. It aims to inform and inspire design teams, where users and stakeholders actively participate in the design process to ensure a good fit between the design and the use of a product. (ID Studio Lab, TU Delft)

Data
The qualitative or quantitative attributes of one or multiple variables. Frequently referred as the lowest level of abstraction from which information and subsequently knowledge are derived.

Design thinking
An iterative, open-ended, holistic and synthesizing way to make sense of a situation, find solutions to complex problems or get unexpected possibilities.

Emergent knowledge
The result of the collective knowledge when put in action.

Empathy
The ability to recognize and share feelings and emotions (such as sadness or happiness) that are being experienced by another sentient being.

Experience
A unique and personal sense making process that happens in a human's mind when consciously interacting with something or someone. It refers to the acknowledgment of the actions according to the stimuli in a determined situation.

Freemium
A business model that consists of offering a free basic version of a service, and subsequently sell premium features on top of that.

Global village
The concept of remotely accessing, sharing and contributing to knowledge and culture from a point in the world to any other location at another point in the world.

Global Positioning System (GPS)
A space-based global navigation satellite system (GNSS) that provides location and time information anywhere on the earth.

Group cognition
A social phenomenon that refers to similar sequences of individual thoughts situated in and emerged out of the interactions of a group. The group can be structured or organized deliberately to produce these thoughts.

Information shadow
The digitally accessible information about an object. (Term coined by Mike Kuniavsky, Smart Things, 2010)

Information
Interpreted data that computers and people can use to make new sense from.

Interaction Design
The practice of designing interactive digital products, environments, systems, and services. Interaction Design also has an interest in aesthetical issues but it's main focus is on human behaviour and on thinking about the impact that particular interactions can have in the individual and social contexts.

Interface Design
The practice of designing devices, software applications, websites or other communication points with the web and ubiquitous technologies. Its aim is to simplify and make as efficient and effective as possible the interaction in order to let the user accomplish his or her goals.

Internet of Things
A world where physical objects are seamlessly integrated into the information network, and where the physical objects can become active participants in processes. Services are available to interact with these 'smart objects' over the Internet, query and change their state and any information associated with them, taking into account security and privacy issues. (Term used by SAP)

Intuition
The ability to build knowledge without inference or the use of reasoned argumentation.

Local village
The concept of accessing, sharing and contributing to the knowledge and culture remotely from any point in the world.

Mashup
A web application that uses and combines data, presentation or functionality from two or more sources to create new functions. The term implies an easy and fast integration, frequently using open APIs and data sources to produce enriched results that were not necessarily the original reason for producing the raw data.

Most Advanced Yet Acceptable (MAYA)
A principle to express that product designs should push technical feasibilities in such a way that they are still accepted by their users in terms of expectations, perception and meaning making.

Meaning
Associative links or relationships in humans' minds between words and objects and experiences that result in the formation of concepts.

Mental model
The way people decode and represent themselves and everything around them.

Meta data
Interpreted data; data about data.

Meta Product
A dedicated network of services, products, people and environments fed by the information flows made possible by the web and other ubiquitous technologies; web-enabled product-service networks.

Near Field Communication (NFC)
A set of wireless technologies that can perform simplified transactions, data exchange, and connections over a short distance.

Network Focused Design (NFD)
Holistic approach to design networks where the focus is on the relationships between services, products, people and environments.

Object augmentation
An object enriched with dynamic information.

Ontology
A representation of knowledge as a set of concepts within a domain, and the relationships between those concepts.

Opportunity cost
The cost of any activity measured in terms of the best previous alternative.

Pervasive computing
A commonly used synonym of ubiquitous computing. It refers to information computing integrated into daily objects and overall activities. Information computing can be both automated or activated by people using interfaces.

Physical computing
Embedding electronics and information processing to objects and materials.

Physical web
The possibility of the web to be embedded in objects and environments adding to them communication and information processing properties.

Product
The result of the industrial process of multiplying material structures with defined functions or purposes in a standardized way.

Product-service network
A dedicated set of relationships between objects, intangible activities, contexts and people.

Proximity
The use of global technology to better fulfill local needs or perform local tasks.

Proximity awareness
The ability of sensing devices to detect the presence of other nearby objects without physical contact.

Quick Response code (QR-code)
A specific matrix barcode that is readable by dedicated QR-code readers and camera telephones. The code consists of black modules arranged in a square pattern on a white background. The information encoded may be text, URLs, or other data.

Reflective practice
The capacity to reflect on action so as to engage in a process of continuous learning. (Termed coined by Donald Schön, 1983)

Resources
In Network Focused Design, all that has perceived value to be exchanged. It could be monetary, time, concepts or knowledge.

Radio-frequency identification (RFID)
A technology that uses radio waves to transfer data between a reader and an electronic tag attached to an object for the purpose of identification and tracking.

Semantic Web
Web of data that facilitates machines to understand the semantics, or meaning, of information on the web. It extends the network of human-readable web pages by inserting machine-readable meta data about pages and how they are related to each other, enabling automated agents to access the web more and perform tasks on behalf of users. (The term was coined by Tim Berners-Lee)

Sensor
A device that measures a physical quantity and converts it into a signal which can be read by an observer or another device.

Serendipity
Phenomenon that refers to the situation when someone finds something unexpected. It occurs when in a particular quest or activity, some elements (a situation, a physical characteristic, and so on) result in an unexpected discovery that usually is regarded as positive.

Service Design
The practice of defining and coordinating a set of activities dedicated to the exchange of valuable resources mainly between people.

Service Dominant Logic (SDL)
A mindset for a unified understanding of the purpose and nature of organizations, markets and society. The foundational proposition of S-D logic is that organizations, markets, and society are fundamentally concerned with the exchange of services — the applications of competences (knowledge and skills) for the benefit of a party. (Created by Vargo and Lusch, 2004)

Service
A set of activities dedicated to the exchange of valuable resources mainly between people.

Smart device
A physical object with embedded artificial intelligence.

Stakeholders
Members of a network of value. They participate in and have influence on the way value is exchanged. Stakeholders could be organizations, customers, companies, or any other entity that is an active participant in a dedicated network of services, products and people.

System mapping
A procedure of defining the flow and function of information and the relationships or connections this information makes possible between elements.

Temporality
A term often used in philosophy that refers to talking about the way time is.

Touchpoint
A communication point where to access, share or manipulate information.

Transdisciplinarity
A research strategy that crosses various disciplinary boundaries to develop a holistic approach towards a complex situation.

Trend mapping
Tracing and categorizing any form of behaviour that appears for a longer period of time and suggests to evolve into a relatively permanent change.

Ubimedia
Digital media (films, news, photos, and social media) accessible in daily activities and a variety of surroundings. (Term coined by Adam Greenfield)

Ubiquitous Computing (Ubicomp)
Information computing integrated in everyday activities and objects in a seamless way.

Ubiquitous technologies
The combined used of two or more technologies where the web is mobile and functions as a stream to enable other technologies (such as NFC, RFID Networks, wireless sensors, memory tags, GPS, printed electronics or augmented reality) to communicate with devices and people.

Web 2.0
A computing platform serving web applications. It is a term associated with applications that facilitate active participatory information sharing and interoperability on the web. A web 2.0 site allows users to interact and collaborate with each other in a social media dialogue as creators of content in a virtual community.

Web 3.0
Web 2.0 added with intelligence for intuitive tasks, customized searches and more interconnections of relevant data for contextual uses. By using ontologies and semantic techniques, web applications in Web 3.0 will 'reason' by themselves to serve personalized functions.

Web
Regularly it refers to the interlinked content accessed via the Internet. However in this book it refers to the Internet together with ubiquitous technologies (see ubiquitous technologies.)

Working memory
The ability to actively hold information in the mind needed to perform complex tasks such as reasoning and learning. Working memory tasks are those that require the goal-oriented active monitoring or manipulation of information.

CHAPTER 1

1. Weizembaum, J., 1976. *Computer Power and Human Reason, From Judgement to Calculation.* San Francisco: W.H. Freeman and Company.
2. Bell, D., 1973. *The Coming of Post-Industrial Society, A Venture in Social Forecasting.* New York: Basic Books.
3. Magerkurth, C., Cheok, A.D., Mandryk, R.L. & Nilsen, T., 2005. Pervasive Games: Bringing Computer Entertainment Back to The Real World. *ACM Computers in Entertainment,* Vol.3, No.3.

CHAPTER 2

1. Juan Enriquez, Chairman and CEO of Biotechonomy LLC. Founding director of the Life Sciences Project at Harvard Business School and a fellow at Harvard's Center for International Affairs. Steward Brand, Founder and curator of Long Now's "Seminars About Long-term Thinking" (SALT), a monthly series of public talks in San Francisco. Katherine Fulton, President of Monitor Institute, Katherine Fulton is also a strategi st, author, teacher and speaker working for social change.
2. Kuniavsky, M., 2010. Smart Things: *Ubiquitous Computing User Experience Design.* Burlington, MA: Morgan Kaufmann.
3. Rich Mogull, founder and CEO of Securosis at the RSA Conference 2011.
4. Kasarda, J. D. & Lindsay, G., *Aerotropolis, The Way We'll Live Next.* Farrar, Straus and Giroux.

CHAPTER 3

1. Suranga Chandratillake, Founder of Blinkx. BBC News, May 2011, Fiona Graham, Technology of business reporter.
2. Johnson-Laird, P.N., 1986. Conditionals and Mental Models. *Conditionals,* p.55–75.
3. Davis, M., 2003. Theoretical Foundations for Experiential Systems Design. *ETP'03 Proceedings of The 2003 ACM SIGMM Workshop on Experiential Telepresence,* p.45–52.
4. Gentner, D. & Stevens, A., 1983. *Mental Models.* Hillsdale, NJ: Erlbaum.
5. Blythe, et al., 2003. *Funology: From Usability to Enjoyment.* Dordrecht: Kluwer.
6. Cacioppo & Patrick, W., 2008. *Loneliness: Human Nature and The Need for Social Connection.* John T. Norton & Company.
7. Theiner, G., 2008. *From Extended Minds to Group Minds: Rethinking The Boundaries of The Mental.* Doctoral dissertation, Department of Philosophy & Cognitive Science Program, Indiana University, Bloomington.
8. Clark, A., & Chalmers, D., 1998. The Extended Mind. *Analysis,* 58, p.7–19.
9. Wasson, C., 2000. Ethnography in The Field of Design. *Human Organization,* p.377–388.
10. Schuler, D. & Namioka, A. (Eds.), 1993. *Participatory Design: Principles and Practices.* Hillsdale, NJ: Lawrence Erlbaum Associates.
11. Sleeswijk Visser, F., et al., 2005. Context Mapping: Experiences from Practice. *CoDesign,* 1 (2), p.119–149.
12. Perkin, H., 1969. *The Origins of Modern English Society, 1780-1880.* Routledge Publisher.
13. Syntens, 2010. *Diensten-innovatie Model, In Zeven Stappen Naar Een Nieuwe Dienst.*
14. Roscam Abbing, E., 2010, *Brand Driven Innovation.* Lausanne, AVA Publishing SA.

CHAPTER 4

1. Shedroff, N., 2011. *Creating Meaningful Experiences.* Online available at: http://www.nathan.com [Accessed June 15, 2011].
2. Pachube, 2011. Online available at: http://www.pachube.com [Accessed June 15, 2011].
3. MacManus, R., 2009. Online available at: http://www.readwriteweb.com/archives/pachube_business_models.php [Accessed June 15, 2011].

4. Klein, J.T., et al., 2001. *Transdisciplinarity: Joint Problem Solving among Science, Technology, and Society.* Basel: Birkhauser.

5. Bearth, T., 2000. Language, Communication and Sustainable Development. *Transdisciplinarity: Joint Problem-Solving among Science, Technology and Society.* Zurich: Haffmans, Sachbuch Verlarg AG. / Häberli, R., et al., 2000. Workbook I: Dialogue Sessions and Idea Market. p.170–175. / Hollaender, K., Loibl, M.C., Wilts, A., 2003. Management of Transdisciplinary Research. Unity of Knowledge in Transdisciplinary Research for Sustainability (Ed. Gertrude Hirsch-Hadorn), *Encyclopedia of Life Support Systems.*

6. Kuniavsky, M., 2010. *Smart Things: Ubiquitous Computing User Experience Design.* Burlington, MA: Morgan Kaufmann.

7. Vargo, S. L., & Lusch, R. F., 2007. Why Service? *J. of Acad. Mark. Sci.,* 2008, 36: p.25–38. / Vargo, S. L., & Lusch, R. F., 2004b. The Four Services Marketing Myths: Remnants from A Manufacturing Model. *Journal of Service Research,* 6, p.324–335.

8. Hanifan, L. J., 1920. The Difficulties of Consolidation. The Consolidated Rural School. *L. W. Rapeer.* New York, Charles Scribner's Sons: p.475–496.

9. Dekker, P., and Uslaner, E. M., 2001. Introduction. *Social Capital and Participation in Everyday Life,* edited by Eric M. Uslaner, p.1–8. London: Routledge. / Uslaner, Eric M., 2001. Volunteering and Social Capital: How Trust and Religion Shape Civic Participation in The United States. *Social Capital and Participation in Everyday Life,* edited by Eric M. Uslaner, p.104–117. London: Routledge.

10. Shirky, C., 2010. *Cognitive Surplus, Creativity and Generosity in A Connected Age.* The Penguin Group.

11. Bleecker, J., 2006. *A Manifesto for Networked Objects — Cohabitating with Pigeons, Arphids and Aibos In The Internet of Things.* Online available at: http://www.nearfuturelaboratory.com/files/WhyThingsMatter.pdf [Accessed June 5, 2011].

12. Felce, D., & Welsh, P., 1995. Quality of Life: Its Definition and Measurement. *Research in Developmental Disabilities.* Centre for Learning Disabilities, Applied Research Unit, College of Medicine, University of Wales, 16, p.51-74.

13. Pariser, E., 2011. *The Filter Bubble, What The Internet is Hiding From You.* New York: Penguin Press.

14. Online available at: http://www.cogknow.eu, http://www.amsterdamlivinglab.nl [Accessed June 15, 2011].

CHAPTER 5

1. Nakielski, K.P., 2005. The Reflective Practitioner. *Decision Making in Midwifery Practice,* Raynor, M.D., Marshall, J.E., Sullivan, A., Elsevier Health Sciences, p.144–145.

2. Schön, D., 1983. *The Reflective Practitioner.* New York: Basic Books.

3. Rolfe, G., Freshwater, D., & Jasper, M., 2001 (eds.) *Critical Reflection for Nursing and The Helping Professions.* Basingstoke, UK: Palgrave. p.26 et seq., p.35.

4. Riding, R., 2000. Cognitive Style: A Review. *International Perspectives on Individual Differences,* R. Riding & S. Rayner (Eds.), vol. 1 Cognitive Styles, p.315–334. Stamford: Ablex.

5. Witkin, H. A., 1962. *Psychological Differentiation.* New York: Wiley. / Guilford, J.P., 1950. Creativity. *American Psychologist,* 5, p.444–454. / Pask, G., 1972. A Fresh Look at Cognition and The Individual. *International Journal of Man-Machine Studies,* 4, p.211–221. / Riding, R., and Cheema, I., 1991. Cognitive Styles: An Overview and Integration. *Educational Psychology,* 11, p.193–126.

6. Kuniavsky, M., 2010. *Smart Things: Ubiquitous Computing User Experience Design.* Burlington, MA: Morgan Kaufmann.

7. Kein, Moon & Hoffman, 2006. Making Sense of Sense Making 1: Alternative Perspectives. *Journal IEEE Intelligent Systems,* 21, 4.

8. Gaver, W.W., Dunne, A., & Pacenti, E., 1999. Cultural Probes. *Interactions VI,* 1, p.21–29.

9. Online available at: http://berglondon.com/projects/magplus/ [Accessed june 20, 2011].

Ford T-Model assembly line, courtesy of Ford Motor Company
RFID chip, taken from Wikimedia Commons
iPod & iTunes as a Meta Product, courtesy of Apple, Inc.
Withings Blood Pressure Monitor, courtesy of Withings SAS
Olinda radio, courtesy of Berg London and BBC
Karotz, by Nivrae (taken from Flickr)
Poken, courtesy of Poken, SA.
Ancient data, by Danny Nicholson (taken from Flickr)
A Social History of Knowledge, by Peter Burke, published by Wiley-Blackwell
Wolframalpha, taken from www.wolframalpha.com
Emotiv EPOC headset, courtesy of Emotiv
Proximity awareness, courtesy of Timo Arnall
Ditto Music, taken from www.dittomusic.com
Aerotropolis, The Way We'll Live Next, by John D. Kasarda and Greg Lindsay, published by Farrar, Straus and Giroux
SenseAware, courtesy of FedEx Corporate Services, Inc.
Innocentive's Global Solver, taken from www.innocentive.com
Current Media, taken from www.current.com
Albert Einstein, courtesy of Smithsonian Institution Libraries
Business Model Generation, by Alexander Ostenwalder and Yves Pigneur, published by John Wiley & Sons, Inc.
Brand-Driven Innovation, by Erik Roscam Abbing, published by AVA Publishing SA
The Echo Nest platform, taken from the.echonest.com
Pachube, taken from www.pachube.com
IBM SmarterPlanet exhibition, courtesy of IBM
Polaroid SX-70 Sonar One Step, by the other Martin Taylor (taken from Flickr)
The Service-Dominant Logic of Marketing, by Robert F. Lusch and Stephen L. Vargo, published by M.E. Sharpe
Paper prototyping web interfaces, courtesy of Booreiland (www.booreiland.nl)
Mosaic thinking, courtesy of Brooklyn Museum Archives (taken from the Goodyear Archival Collection)
Smart Things, Ubiquitous Computing User Experience Design, by Mike Kuniavsky, published by Morgan Kaufmann
Cultural probes, courtesy of Muzus (www.muzus.nl)
Role playing, courtesy of Muzus (www.muzus.nl)
Personas, courtesy of Muzus (www.muzus.nl)
Visual Thinking, courtesy of Jam Visual Thinking (www.jam-site.nl)
Creative session, courtesy of Zilver Innovation (www.zilverinnovation.com)
Co-creation session, courtesy of Fronteer Strategy (www.fronteerstrategy.com)
Storyboards, courtesy of Booreiland (www.booreiland.nl)
Magplus, courtesy of Berg London and Bonnier R&D
Network Focused Design session, courtesy of Booreiland (www.booreiland.nl)

We would like to thank all the collaborators of this book: Jan Buijs, Federico Casalegno, Jenny de Boer, Alexandra Deschamps-Sonsino, Harald Dunnink, Marc Fonteijn, Maddy Janse, Sebastian Kersten, Lucy Kimbell, Anne Lise Kjaer, José Laan, Frido Smulders, Anna Valtonen and Jeroen van Geel. Every conversation and every insight was inspiring and helpful for the realization of this book.

Also thanks to Rudolf van Wezel, owner and CEO of BIS Publishers and the Creative Company Conference. Without his drive for innovative publications and his confidence in us, this book wouldn't have been possible.

Thanks to Booreiland's graphic designer Joost Krijnen, for his bracing contribution and his dedication beyond the call of duty.

Thanks to Mike Kuniavsky, for writing a most energetic foreword. His words convey the essence of this book in an inviting way.

Thanks to the guidance and enthusiasm of dr.ir. J.A. Buijs and ir. E. Roscam Abbing when mentoring Sara during her master's thesis project. Her thesis set the academic background of the Network Focused Design approach.

Thanks to our co-bloggers who shared their insights and experiences on Metaproducts.nl: Tal Benisty, Jurjen Helmus, Jacco Lammers, João Rocha, Anne Marleen Olthof, Stefan Veen, and Jelmer Zijlstra.

Thanks to our design studio friends Jurjen Helmus, Cristel Lit, Jetze van Beijma, Jornt van Dijk, and Jelmer Zijlstra for their support and all the laughs. Cheers guys!

Last but not least, thanks to all our friends who have become loyal believers and supporters of the Meta Products topic.

A

B

C

D

E

F

G

H

I

J

K

L

M

N

O

P

Q

R

S

T

U

V

W

*Visit **metaproducts.nl** to find out more about Meta Products and Network Focused Design*